SOLDIERS, COMMISSARS,
AND CHAPLAINS

SOLDIERS, COMMISSARS, AND CHAPLAINS

Civil-Military Relations since Cromwell

Dale R. Herspring

ROWMAN & LITTLEFIELD PUBLISHERS, INC.

Lanham • Boulder • New York • Oxford

ROWMAN & LITTLEFIELD PUBLISHERS, INC.

Published in the United States of America
by Rowman & Littlefield Publishers, Inc.
4720 Boston Way, Lanham, Maryland 20706
www.rowmanlittlefield.com

12 Hid's Copse Road, Cumnor Hill, Oxford OX2 9JJ, England

British Library Cataloguing in Publication Information Available

Library of Congress Cataloging-in-Publication Data

Herspring, Dale R. (Dale Roy)
   Soldiers, commissars, and chaplains : civil–military relations since Cromwell / Dale R. Herspring.
     p. cm.
   Includes bibliographical references and index.
   ISBN 0-7425-1105-7 (cloth : alk. paper)—ISBN 0-7425-1106-5 (paper : alk. paper)
     1. Chaplains, Military—History. 2. Morale. 3. Armed Forces—Social services. 4.
   Communist countries—Armed Forces—Political activity. 5. Civil–military
   relations—History. 6. Civil supremacy over the military—History. 7. Sociology, Military.
   I. Title.

UH20.H47 2001
355.3'47'09—dc21   2001016214

♾™ The paper used in this publication meets the minimum requirements of American
National Standard for Information Sciences—Permanence of Paper for Printed Library
Materials, ANSI/NISO Z39.48-1992.

To Laurie Bagby and Jim Franke

I would as soon go into battle without my artillery than without my chaplain.

—Field Marshall Bernard Montgomery

A political officer in combat. . . . He works to . . . raise the spirit of the troops in the attack, to ensure that the soldiers are fed on time, to fight for a high level of discipline, to keep them moving forward, to carry out reconnaissance. . . .

—E. E. Mal'tsev, V gody ispytaniy

Contents

Preface

THE IDEA FOR THIS BOOK goes back some thirty years, to a time when I was a chaplain's assistant in the U.S. Navy. I was fascinated by the work that chaplains did—especially since I had a front row seat. And this work included not only the religious aspect of the job, although that was certainly important to the men I assisted. Chaplains also spent countless hours counseling sailors and their families, attempting to resolve seemingly intractable problems. I can still remember an instance when the Shore Patrol brought in a sailor who had beaten his Japanese-born wife and threatened her with deportation if she told anyone. The chaplain did something that is very uncharacteristic of those in his profession. Instead of waiting for the commanding officer to do so, he personally preferred criminal charges against the young man in question and told the Shore Patrol to lock him up. The chaplain's God was a loving one but also a very just one.

The chaplains spent a lot of their time attempting to motivate sailors—to convince them that they could do better and that the country they served was worthy of the sacrifices they might be called upon to make. When it came to questions of patriotism, chaplains were not neutral.

When I left the Navy and entered college, I decided to specialize in European affairs—especially those of Germany, Eastern Europe, and the former USSR. I soon forgot about my experience with chaplains. Later, however, I found myself dealing repeatedly with the former Soviet military, both as a policymaker in the Department of State and as an academic. One person who continually caught my eye was the Soviet political officer. I was told by many of my colleagues that he was a commissar and spent most of his time making sure that regular officers behaved themselves.

Later, I had the fortune to meet a man who radically changed my view of political officers. I was stationed at the American Embassy in Warsaw, Poland, and had spent several weekends working on an article dealing with political officers in the Polish Army. I gave Professor Jerzy Wiatr, who had recently taught at that country's Political-Military Academy, a copy of my draft article. Seldom have I received such helpful, if unwelcome at the time, comments. "You are describing the situation in the very early years of this army. Political officers are not the same as political commissars. You need to start over and do a more realistic analysis of the problem. Political officers are not primarily there as controllers, as you seem to think. Some of that goes on, but they are critical. They help make the military function effectively." Needless to say, I took Professor Wiatr's advice, and upon further investigation, it turned out he was right. This led me to question the comments my colleagues had made about Soviet political officers. Indeed, the more I looked into the matter, the more I began to realize that Professor Wiatr was right about the contemporary Soviet political officer as well. The Soviet political officer's primary concerns seemed to be the same things that motivated chaplains—morale, motivation, counseling, and resolving political socialization.

The similarity of the Soviet political officer to the American chaplain was brought home to me during a visit by U.S. ships to Vladivostok in 1989. I remember talking to a young Soviet political officer—asking him what he planned to do if communism collapsed in Russia. I will never forget his response. "That is not a problem. I am just like your chaplains (*Kapellani*)." I looked at him somewhat bewildered, noting, "But you are an atheist!" He answered, "That is not a problem. When it comes to our day-to-day work, we do the same thing."

When I thought about what my Soviet colleague had said and what I had seen of the Soviet armed forces over the years, I began to realize the truth of his observation. Despite his attempt to make me believe that religion didn't matter when it came to his job description, I decided that he was right—Soviet political officers and American chaplains probably did have more in common than most Westerners realized.

So when the opportunity presented itself, I decided to take a closer look at the issue in a systematic fashion. I wanted to know why political officers existed and how their roles changed over time. I also wanted to know how their functions differed or were similar to those of an American chaplain. I believed that this was crucial in helping people understand the changing nature of civil-military relations in party states such as the former USSR.

The more I thought about the issue, however, the more I became convinced that such a study should go beyond the American and Soviet cases. It should be constructed to deal with change. As a policymaker, I had always been bothered by the failure of the role of the Soviet political officer never seemed to change for certain people in the intelligence field. To overcome this problem, I decided that it would be essential to include in the study other historical precedents (e.g., the influence of the French Revolution's commissar on the Soviet commissar or the Soviet commissar on the Soviet and East German political officers). My hope was that these cases would shed light on the development of both the chaplaincy and political officers. I was primarily interested in determining to what degree similarities existed between the various experiences and what such an analysis would tell us about civil-military relations in party states. Could we make some general statements about the nature of these officers and the kind of political systems they served? The result is this book.

Acknowledgments

A NUMBER OF PEOPLE PLAYED an important role in helping me write this book. First, and foremost, were the members of Kansas State's Interlibrary Loan Department. I can't count the number of times the women who work there went out of their way to assist me in finding difficult-to-locate books and articles in a variety of languages. I am very much in their debt.

I am also indebted to Colonel Hans-Werner Weber, formerly of the East German Army, who assisted me in obtaining equally difficult-to-locate sources on the origins of the East German military and the role played by political officers in the NVA. As in the past, Colonel Weber's help was crucial when it came to issues related to the former National People's Army. I would likewise like to thank Professor Jerzy Wiatr of Poland for his continuing helpful insights on the question of political officers in communist militaries.

On the American side, I am most indebted to my long-time friend and faithful critic Dr. Jake Kipp. One of the problems with having friends read your work is that they often provide more positive evaluations than the material warrants. A real friend is one who is prepared to tell you when he or she believes your work can be improved. Fortunately for me, Jake Kipp is that kind of person, and I am deeply in debt to him for taking the time to critique the entire manuscript.

As was the case with my last book, I was also incredibly lucky to have Susan McEachern of Rowman & Littlefield serve as my editor. Her unfailing efforts to teach me the fundamentals of English, while at the same time making suggestions on how to improve the book, were greatly appreciated. I would also like to thank my colleague Kisangani Emizet for reading several chapters. His insightful eye enabled me to avoid some of the conceptual pitfalls to which a book like this is prone. Scott Tollefson also supplied many useful thoughts on the role played by militaries in Latin America. Any mistakes or omissions are obviously my responsibility.

I would be remiss if I did not thank my wife and youngest son, who have patiently listened to more discussions about political commissars, political officers, and chaplains than they ever wanted to. Their encouragement and patience meant a lot to me.

Finally, this book is dedicated to two of my colleagues, Dr. Laurie Bagby and Dr. James Franke, both of whom stood by me in a very difficult situation. Their encouragement and support meant more than they will ever know.

Introduction

From time immemorial it has been recognized that religion, and religious faith, was a powerful factor in helping men to fight nobly for their country; it was important that they should believe in the justice of their cause and that they should feel that God was on their side.

—John Smyth, *In This Sign Conquer*

Understanding the Role of Political Commissars

ONE OF THE MOST FRUSTRATING experiences of the Cold War for me, both as a policymaker and as an academic dealing with communist militaries, was the "closed" mindset I encountered among other analysts, policymakers, and academics. The majority seemed to believe that the role of the party within these militaries was both unique and immutable. While working with these armies—including my rather extensive contacts with officers in communist armed forces—I began to realize that this was a gross simplification.

There was no question that the nature of civil-military relations in communist polities—and especially the role of the party in the army—was different than in Western-style democratic systems. But unique . . . ? As I studied these militaries, I became convinced that certain commonalities existed between East and West. And here I am talking not only about uniforms, weapons, doctrine, or customs. Most analysts would have accepted the premise that there were important similarities in these areas. However, analysts often found it difficult to agree that similarities existed in what might be considered politically sensitive areas as well. For example, the idea that communist political officers and noncommunist chaplains carried out similar functions was considered by many to be blasphemy. How could anyone possibly suggest that the two roles were in any way alike? Yet as I studied these militaries, I became convinced that was exactly the case. While there were important differences, the roles played by chaplains and political officers were very similar when it came to issues of morale, counseling, and political socialization.

As far as change was concerned, I was struck by the tendency of many in the field to refer to communist political officers as commissars. While this was certainly true of the civil war period, it didn't seem to make a lot of sense as I spoke with officers from communist militaries. It soon became obvious that there were important differences: a commissar was a control device,[1] whereas a political officer was only involved in control in a very peripheral sense. However, if the role of these officers changed over time, this suggested that the relationship between the military and the party was also changing, a development that had not only academic significance but important policy implications as well.

In writing this book I also discovered that some observers confused chaplains and commissars. In Jonathan Adelman's book on revolutionary armies, for example, he referred to chaplains in Cromwell's armies as "the seventeenth-century version of modern commissars."[2] In fact, nothing could be further from the truth. To be sure, certain common functions were carried out by both Cromwell's chaplains and commissars in the Russian military, but some very significant differences existed; for example, chaplains in Cromwell's army never carried out a control function.

I hope to correct these misconceptions, which developed during the Cold War, and in the process provide some insights into the functions and roles of these vitally important institutions.

Dealing with Change

There were good reasons for our failure to look closely at change in communist (or even totalitarian) regimes during the Cold War. To begin with, these societies were "closed" in the sense that they made a conscious effort to hide from public view what was going on—especially in the military sphere. The definition of what was secret in these systems went far beyond Western norms. Getting the most basic information—even if one had access to classified information—was very difficult.

Coupled with this lack of basic information was a tendency on the part of many scholars to assume that these ideological systems were consumed with control, in an effort to ensure that the political center's hegemony over all parts of the political system was total. This idea was evident in the pioneering work by Carl Friedrich and Zbigniew Brzezinski. These renowned scholars believed that the Communist Party aimed to limit any sign of autonomy in the military.[3] As a consequence, the party elite made certain that the military was totally subservient to the Communist Party and, to that end, developed a number of control devices—one of which was the position of political commissar.

Since the work of Friedrich and Brzezinski, several scholars have suggested a variety of other approaches, but as I have pointed out elsewhere,[4] none dealt effectively with change in communist or totalitarian systems. One (which was based on the Friedrich and Brzezinski approach) effectively described civil-military relations when the interaction was characterized by conflict,[5] while another did a good job of analyzing the situation when the party was almost totally dominant.[6] A third approach focused on policy issues, and while this marked a step forward in dealing with change, by noting the changing role the military played in Soviet politics, it did not explain *why* change occurs.[7] Finally, the author of this theory made an attempt to look at Soviet military politics by focusing on the senior military leadership and its interaction with party officials, but he, too, never really dealt with the question of change.[8] Almost all of the other models have relied on these paradigms, especially the conflict model proposed by Kolkowicz.

If any lesson can be learned from our trying to understand civil-military relations in a party-state, it is that change constantly takes place. The changes may not be as obvious as they are in open systems, but they occur nevertheless. The task for the analyst is to learn how to detect and explain change. Thus, I argue that paradigms that are unable to deal with change over time do little to expand our understanding of civil-military relations in such systems.

It is important to emphasize that these closed systems are influenced by their political culture and do not change at the same pace or in the same way. However, one major factor that helps explain change in civil-military relations in party-states is the issue of value congruence. When there is a clear difference in values between the country's rulers and its military elite, the situation will be marked by conflict. This is not normally a problem in democratizing states, where the primary task is to integrate the military into a democratic system. Conversely, in totalitarian states the first task of a political elite when it comes to power is to develop control devices to limit military autonomy, thereby ensuring that the armed forces remain politically reliable even if this means undermining efficiency. At the same time, most successful party leaderships will be pragmatic enough to make some concessions to the military if the regime is under serious attack. The degree of independence depends on how important the regime's goals are to the political leadership and how critical the military is to achieving those goals.

Assuming that the regime survives and stabilizes, over time a value congruence will develop between the political leadership and its military. Those who do not share the leadership's vision will either choose to go elsewhere or they will be purged. Gradually, a new generation of officers and noncommissioned officers (NCOs) that share the values of the regime will take over these positions. As conflict between the military elite and the political elite declines, so, too, does the need for the old control devices. As a result, assuming the regime is not under serious internal or external threat and value congruence between the military and civilian leaderships continues to converge, the regime will grant the military ever-increasing degrees of autonomy.

Functional Similarities

This brings us to our second thesis: namely, when the congruence of values between the military leadership and the political leadership has reached the point where civilians no longer see the military as a threat, the institutions previously utilized to "control" the military will begin to limit themselves to areas such as motivation, morale, and political socialization.

Furthermore, it is my view that these new functions parallel those in other militaries. Indeed, they are tasks that must be carried out in any military at any time in history, regardless of the country's political system. For an army to be effective, all three functions of motivation, morale, and political socialization must be carried out. Without good morale or a strong sense of motivation, few armies will emerge successful on the battlefield. Political socialization, too, is crucial. Without it, the cohesion that is so vital to military success will be undermined.

Historically, one of the most successful vehicles for ensuring a high level of motivation has been religion—both the ceremonies associated with it and the belief that, by fighting, the individual soldier increases his chances of an enjoyable life in the hereafter. In the Roman army, for example, religious ceremonies were held prior to battles, and while it is difficult to determine whether these religious events led soldiers to fight better in battle, it is worth noting that a belief at that time posited that the "souls of soldiers killed in battle whilst displaying heroism received better treatment in the afterlife than those of men who died of old age or disease."[9] The one study that looked at this issue in detail argued that "these practices served to identify the soldier and his unit with Rome's destiny."[10]

While it has always been possible to assign to regular or line officers such tasks as building morale, motivation, and socialization among soldiers, throughout history the military has tended to use specialists for these tasks. For example, the Bible tells of the presence of priests during Joshua's battle for the town of Jericho,[11] while at various times the Roman army had chaplains or spiritual advisers who helped with such matters.[12] Clearly, issues such as motivation, morale, and socialization were important even in antiquity.

Chaplains also served in Charlemagne's army and during the Middle Ages, most particularly during the Crusades. According to one source, an order issued during the Second Crusade stated:

> Every ship should have its own priest, and that there should be orders to observe the same practices as in a parish. . . . That everyone should confess weekly and go to communion on the Lord's Day.[13]

In addition to the liturgical functions noted here, chaplains also helped the soldiery by hearing confessions, providing comfort to crusaders, and giving all parties a religious understanding of the reasons for the Crusades.[14] During the Reformation, chaplains worked primarily "to keep the individual and the army close to God," an approach that will become clear when we look at the situation under Cromwell.[15]

Aside from their distinct religious duties, chaplains were further distinguished from other officers by their noncombatant status. This can be traced to the Jesuits during the Spanish conquest of the New World. As Washburn put it, "Possibly the greatest contribution of the Jesuits to the history of the chaplains is their demand that he shall not bear arms," at a time when it was common for chaplains to carry weapons just like any other soldier.[16]

I mention the Romans and the Crusades to demonstrate that these specialized functions were being carried out long before any of the more modern militaries emerged. Morale, motivation, and political socialization are integral parts of any military, although—as I argue later—needs will be greater in militaries that either rely on conscripts or depend on a religious or ideological framework to motivate soldiers.

Motivation

The common denominator in all seven cases analyzed in this book is that, with the exception of Cromwell's army,[17] civilian authorities were dealing for the first time with mass militaries. The mercenaries who had staffed most armies still operated, but a new kind of soldier—the conscript—was emerging. In the professional militaries that preceded the introduction of ideologically driven armies during the English Civil War in the late fifteenth century, plunder and pay served as motivating factors. Ideology was irrelevant.

However, with the rise of what might be called the ideological state, or with the beginning of religious wars, the state and its generals were faced with a new task: motivating "citizen" soldiers to the point where they were prepared to lay down their lives for an ideal, a concept, or a religious end, regardless of whether it was for God, "Liberty and Equality," the Proletariat, the Führer, proletarian internationalism, or freedom and democracy. Such an undertaking would be far from easy. In most cases, pay was low and the incentive to join the armed forces nonexistent. In a nutshell, one had to find a way to make the soldier or sailor "want" to die for his country or revolution.

Political Socialization

Since many of these regimes were revolutionary—they were replacing not only a regime, but a whole political system as well—it was also necessary to change values. Political socialization was therefore crucial. The new leaders of the French Revolution had to convince large numbers of often illiterate individuals to support a system based on totally new and different premises. In France, liberty and equality were substituted for God and king. How to ensure that the bonds tying soldiers to the ancien régime were broken and their place taken by a sense of commitment to a new one? The same was true of the Russian Revolution. How to break the commitment to God and czar and replace it with one to the revolution and the working class?

This task would not end when most soldiers and officers had accepted the regime's basic value system. It would have to be constantly reinforced. Soldiers and sailors would have to be convinced over and over that their regime was the best available and that when it came to individual behavior, vices such as alcohol and drug use were wrong and a strong sense of discipline was in their own self-interest. This continuing need for socialization would be evident in Cromwell's army, the Wehrmacht in World War II, or even in the American military, not to mention the Soviet or East German militaries.

Morale

Morale was a key concern of all military—and political—leaders. Someone had to be around to ensure that basic needs were taken care of. While a commander was responsible for such matters, someone else was usually assigned the task of ensuring that food was satisfactory, that lodging was appropriate, and, most important, that personal problems were dealt with in an effective manner.

Identifying the Actors

Three different kinds of morale, motivation, and political socialization officers are described in this book. We now turn to a definition of the roles played by each.

As noted earlier, the key issue facing regimes in which a significant value divergence existed between the military and the political elite was to find a way to ensure political control of the military while also taking care of concerns such as motivation, morale, and political socialization.

Toward this end, political authorities in France, in the Soviet Union, and, to a lesser degree, in Hitler's Germany relied upon what have been called *political commissars*. *Webster's Third New International Dictionary* notes that the word *commissar* comes from the Middle Latin—*commissarius*—"one to whom something is entrusted." This is exactly how the French and Russian revolutionaries looked upon the commissar. He was given the task of ensuring that the line officers he dealt with were reliable. This was especially true of the commanding officer with whom he worked. Unless the commanding officer was reliable, he could undermine everything the regime was fighting for. This is why the commissar was often put in a position of authority (the French Revolution) or shared authority with the commander (the Russian Revolution). Control was his primary task.

However, this individual was also expected to make certain that the unit's combat readiness was as high as possible, as well as to undertake the political socialization of both soldiers and officers. Too often the soldiers participating in the revolution had no idea why they were fighting. Ideas such as "Equality" or "The Working Class" often meant nothing to them. Yet to convince common soldiers to fight and sacrifice their lives, it was essential to inculcate these terms and their values. Needless to say, this was a daunting task. Morale and motivation were also major concerns.

Although the political officer was closely related to the commissar and in fact evolved from him in both the Soviet and East German cases, he carried out quite different functions. To begin with, in these two instances he was no longer considered the equal of the commander. He was clearly subordinate to him. The political leadership had made a conscious decision to move away from a political commissar to a political officer. He had an alternative chain of command in the party structure through which he could report negative actions on the part of the commander. However, every political officer clearly understood that his fate was closely tied to that of the commander. If the unit failed in its training exercises, or if an insufficient number of soldiers or sailors qualified in their military specialty, the political officer would be considered just as guilty as the commander. As a result, he had to think very carefully before taking the extreme step of reporting his commander to higher political officials. It could very easily backfire.

It is worth noting that although unusual, it was also possible for the system to move from having a political officer back to having a commissar. This happened in the Soviet Union on two occasions. While there was no clear terminological change in the World War II Wehrmacht political officer, this also happened at the end of World War II, with the tendency of National Socialist Leadership Officers to assume greater degrees of control over line officers.

The political officer was also responsible for motivating soldiers and sailors. If they were not performing at the appropriate level, it was a political problem, and since the political officer was the deputy for political work, he was responsible. The same was true of political indoctrination. If soldiers did poorly on their political exams, the political officer would have some explaining to do.

Finally, the political officer also dealt with morale problems. If an individual had a drinking problem, it was up to the political officer to see that this person received help. If a soldier had problems at home, he was encouraged to discuss the matter with his political officer. Similarly, the political officer was expected to worry about things like the soldier's housing, his meals, or how he was treated. In fact, this often took up a major part of the political officer's time. As a former East German officer put it, "The only time I dealt with a political officer outside of political lectures was when there were personnel problems."[18] In essence, the political officer had become more of a morale, motivational, and socializing officer than a controller.

The military chaplain goes back to the fourth century, much further back than either the commissar or the political officer. St. Martin of Tours was a French soldier who purportedly met a shivering beggar on a cold winter night. Having no money, he cut his cloak in two and gave half to the beggar. After a vision, he left the army to devote his life to the Church. He soon became the patron saint of medieval French kings. His cloak (*cappella*) was considered a sacred relic and was

TABLE 1
Functions of Chaplains, Political Commissars, and Political Officers

	Provide Spiritual Comfort	Motivate, Counsel Soldiers	Political Socialization	Change Basic Value System	Control Actions of Line Officers	Assume Command
Chaplains	X	X				
Political Commissars		X	X	X	X	
Political Officers		X	X			

carried into battle by a priest who was called a *cappellanus,* as a means of motivating the troops. It is from the latter word that the term *chaplain* is derived.[19]

The chaplain occupies a unique position, whether in Cromwell's army or the American military. He "is not just half-military and half-church. He is fully a member of both institutions. Though he leaves the job environment of the church, he retains his full institutional status."[20] As the chapter on the American chaplain will demonstrate, his role has changed over the years. Nevertheless, two aspects of his job have remained constant. The first is his obligation to provide spiritual advice and comfort, and the second is his responsibility for motivation, morale, and, to a certain degree, political socialization. In this sense, his job is similar to that carried out by the political officer—although, as the reader will note, there were also important differences.

Table 1 illustrates the relationship among chaplains, political officers, and political commissars insofar as their functions are concerned. This table shows that the distinction between each officer is not as clear as one might wish. None of the categories are exclusive. All of them overlap to a degree. Chaplains were involved in motivation just as political officers were. They also provided spiritual comfort, something that political officers did not do—*unless one considers preaching a secular ideological doctrine to be a spiritual undertaking.*

Political commissars assumed all of the roles noted previously, including the control function, although they did not give spiritual advice. The only area where Soviet political commissars were not active was in assuming command. They shared command with the commander and on occasions they might be expected to assume command if other senior officers in a combat unit had suffered serious casualties, but, based on the Soviet example, they seldom did so. As the reader will note, the French case was somewhat different and is an exception to the rule.

Structure of the Book

For reasons of space and because certain experiences more clearly show the interaction of civilian and military/political authorities than others, I used seven historical incidents as case studies. Others, such as the Chinese Communist revolution, the World War II Yugoslav experience, or Ghana's regime under Kwame N'krumah, could have been utilized, but given the importance of the seven regimes I selected, as well as the large amount of secondary material available on them, I limited the focus of my study to these cases.

The book begins by tracing the historical evolution of the chaplain in the American armed forces. The reason for putting the American chaplain first is simple; it is the basis against which all of the other institutions will be compared. The American chaplain never engaged in "control" nor did he attempt to impart a new, ideological system to the troops he served. To be sure, his role changed throughout the 225 years of his existence, but the one thing that remained constant was his concern with morale, motivation, and political socialization.

With the U.S. experience as a backdrop, the rest of the book is devoted to an analysis of six other cases. They include the English Civil War, the French Revolution, the Russian Revolution, the German Wehrmacht in World War II, the post–World War II Soviet Army, and the East German experience during the Cold War.

The English Civil War (1642–1647)

The English Civil War represents one of the first conflicts in which nonprofessional soldiers played a major role. Furthermore, it also was one of the first cases in the last five hundred years where political authorities relied heavily on special individuals—in this case, chaplains—to motivate and inspire soldiers prior to battle and during the conflict and to censure or discipline those who did not perform satisfactorily after the battle was over.

The French Revolution (1789–1799)

The French Revolution launched the use of mass armies. It also represents one of the first instances when a new regime attempted to introduce a new (and in this case, secularized) ideology at a time when much of the population remained tied to old beliefs. This created a situation where political authorities had to come up with a device that would resocialize large numbers of soldiers, and they also faced the difficult task of controlling the actions of many professionals from the old regime. In many instances, these latter individuals were not only potentially disloyal, but, as the defection of General Dumouriez during the French Revolution illustrated very clearly, a number even went over to the other side. As a result, political authorities in France employed both political commissars and representatives on mission. The two were utilized to ensure the political reliability of the officer corps, as well as to resocialize the mass of French soldiers.

The Russian Revolution (1917–1930)

For most people in the West, the term *political commissar* is synonymous with the Russian Revolution. The reason is simple: the events took place in the not-too-distant past and they have been heavily publicized in the West. The need for political commissars was great. As Lenin pointed out on a number of occasions, were it not for the "military specialists" from the czarist regime, the Red Army would have had little chance of defeating the Whites during the civil war. Communist commissars—most of whom often knew little or nothing about military matters—were convinced Marxists-Leninists. They were given equal authority with the military specialists and expected to not only make certain that the latter behaved themselves but to ensure that the soldiers were prepared to give their lives for the revolution.

National Socialist Leadership Officers (1933–1945)

The situation facing Hitler and the Wehrmacht during World War II was different from both the French and Soviet cases, in that it was not a question of changing soldiers' deeply held values—although many of the country's senior officers had serious questions about Hitler and his intentions. In the German case, the key problem was one of maintaining military cohesion and efficiency at a time when the Wehrmacht was suffering one defeat after another. How to make certain that military personnel would not adopt a defeatist attitude? How to ensure that they would fight to the last man at a time when fighting made little sense—since it was becoming clear to everyone that the war was lost?

The answer came in 1943 when National Socialist Leadership Officers were introduced in the Wehrmacht. There would be a major battle between the Nazi Party leadership and the country's generals over the exact role these officers were to play. The party leadership wanted them to assume the role of a commissar, while the generals did everything possible to keep their function limited to that of a political officer for fear that the NSFOs would interfere with the commander's ability to command his unit. In the end, they would begin to resemble political commissars, but by that time the war was lost and most members of the Wehrmacht were in prison camps.

Post–World War II Soviet Political Officers (1945–1990)

The actions of political officers in the Soviet military in the aftermath of World War II demonstrate clearly how circumscribed their authority was. They continued to motivate soldiers, worry about morale, and take charge of political indoctrination lectures. However, their authority vis-à-vis line officers and especially the commander was extremely limited. The commander was very much in charge; the political officer was one—and only one—of his deputies. Furthermore, he almost never assumed command except in the most unusual cases. The political officer filled out annual reports on the political reliability of officers and NCOs, but such reports were generally routine.

As the Soviet regime moved toward collapse, the High Command relied more and more on political officers as a means of ensuring military cohesion. Nationality disputes were one of many factors that tore the military apart, and it was up to a frequently revamped political apparatus to try and patch it together. This is why the Soviet High Command fought so vehemently against the dissolution of the party's preferential position within the country. When the party structure disappeared, so, too, did political officers. The result further undermined the army's cohesion.

East German Political Officers (1945–1990)

The East German case, like the situation in other East European armies, was very different from the Soviet one in 1917. The main reason was that these regimes, and the armies that supported them, came to power on the back of the Red Army. As a consequence, while East Germany had political commissars (called P-K—politicalcultural officers), they were in place for a very short period of time as a means of keeping an eye on the former members of the Wehrmacht, who were taken into the armed forces because of the lack of military expertise on the part of German communists. Soviet advisers, who were very prolific in number during the early days, as well as the presence of the Soviet armed forces (which were stationed in their country), obviated the need for such a control device over the long run.

Furthermore, when it came to the GDR, the vast majority of those who were selected to attend officers' schools came from the lower classes and recognized that if it were not for the new communist regime, they would never have had an opportunity to move up the socioeconomic ladder. As a result, their political reliability was ensured almost from the start.

Methodological Issues

This study does not intend to develop a general theory of civil-military relations. It is my opinion that we are a long way from developing such a theory. In a certain sense, we are like a small child—still crawling. It will be some time before we begin to walk, much less run. Having said that, I think that we can and should still look for commonalities and differences between political systems over time. What actions appear to be common between systems? In which cases does political culture appear to play a major role?

My biggest concern, however, is with change in civil-military relations, and particularly those communist or other totalitarian systems. Are there general statements that we can make with regard to civil-military relations in party-states? Is it true that party regimes will only permit autonomy within the military if the regimes are convinced that a value congruence exists between the officer corps and the political system? What about periods of instability? At what point is the regime likely to crack down on the military and introduce intrusive political control devices like political commissars? Is there a degree of autonomy that is crucial for any military to operate? Why do political officers and chaplains intersect insofar as their jobs are concerned, when it comes to morale, motivation, and political socialization problems?

Assuming we can establish some relationships between these regimes, can they be generalized with regard to autocratic or aristocratic regimes? Why was it not necessary for the United States or for Cromwell to introduce political officers, not to mention political commissars? Why was the United States able to handle the instability and stress of World War II better than Germany? Was it just because the Germans were losing, or was a bigger issue at stake? These are only a few of the questions that will be discussed throughout this study.

Another general question is, what do these experiences tell us about the overall issue of civil-military relations? To what degree does the role played by these officers (e.g., commissars, chaplains, or political officers) reflect the country's political culture or the nature of its political institutions? Is there an inevitable need for such officers in every military? The fact that democratic, authoritarian, and totalitarian militaries seem to have them suggests that an officer of this kind is a prerequisite, regardless of the nature of the country's culture or political system. The difference lies in function. Some are controllers, while all are concerned with motivation, morale, and political socialization. Why? This book will attempt to answer that question.

Sources

Writing a book that deals with such diverse subjects presents a challenge to any writer. Fortunately, when it came to the different cases I examined, I quickly discovered that others had done most of the basic research.

For anyone interested in studying the evolution of the American chaplain, the Army's chief of chaplains has prepared an in-depth study of the issue from the early colonial period to the present. Other sources are available, but this is one of the most helpful.

The English Civil War has been the subject of countless books and articles. Among the most notable is Christopher Hill's *God's Englishman: Oliver Cromwell and the English Revolution*.[21] Other important and useful books include Leo Solt, *Saints in Arms;* Ian Gentles, *The New Model Army;* Charles Firth, *Cromwell's Army;* and Ann Laurence, *Parliamentary Army Chaplains, 1642–1651.*

A number of books have been written on the role of the revolutionary army during the French Revolution. Among the more important are John A. Lynn, *Bay-*

onets of the Republic; Alan Forrest, *Soldiers of the French Revolution;* Jean Paul Bertaud, *The Army of the French Revolution;* and William S. Cormack, *Revolution and Political Conflict in the French Navy, 1789–1794.*

The Russian Revolution has been written about by myriad writers. As far as the role of political commissars is concerned, the best study in English is Mark von Hagen's *Soldiers in the Proletarian Dictatorship,* although Dmitri Fedotoff-White's *The Growth of the Red Army* and John Erickson's *The Soviet High Command* are also valuable. Regarding Russian sources, the most important are: I. B. Berkhin, *Military Reform in the USSR, 1924–1925;* A. G. Kavtaradze, *Military Specialists in the Service of the Republics of the Soviets, 1917–1922;* and the classic Soviet study of political officers, Yu. Petrov, *The Development of Political Organs, Party and Komsomol Organizations in the Army and Navy.*

As much as I can determine, the only material available in English on the NSFOs is in the form of a doctoral dissertation: Robert Lee Quinnett, "Hitler's Political Officers: The National Socialist Leadership Officers," University of Oklahoma, 1973. The key studies in German are: "Adolf Hitler and the NS-Leadership Officer (NSFO)"; "On the History of the National Socialist Leadership Officer (NSFO)"; Volker R. Berghan, "The NSDAP and 'Spiritual Leadership' of the Wehrmacht, 1939–1943"; and Arne W. G. Zoepf, *The Wehrmacht between Tradition and Ideology.*

Concerning the postwar Soviet political officer, I relied heavily on books such as Alexander Khmel, *Party-Political Work in the Soviet Armed Forces,* and Petrov's *The Creation of Political Organs, and Party and Komsomol Organizations in the Army and Navy.*

The information in this book about East German political officers relies heavily on interviews conducted by me, as well as on memoirs from former East German officers. Some of the more noteworthy of the latter include: Werner Rothe, *Years in Peace: A GDR Biography;* Erich Hasemann, *Soldier of the GDR;* as well as the interviews that Jürgen Eike prepared for his television documentary "The Disappearing Army."

Notes

1. By the word *control* I have in mind a situation in which civil authorities assume that the military is potentially (or really) hostile and create devices to ensure that it remains reliable because its use is crucial to regime survival.
2. Jonathan Adelman, *Revolution, Armies and War* (Boulder, Colo.: Riener, 1985), 26.
3. See Carl J. Friedrich and Zbigniew K. Brzezinski, *Totalitarian Dictatorship and Autocracy* (New York: Praeger, 1965), 10. This book was originally published in 1956. Brzezinski's ideas vis-à-vis this topic are contained in his *Political Controls in the Soviet Army* (New York: Research Program on the USSR, 1954).
4. Dale R. Herspring, *Russian Civil-Military Relations* (Bloomington: Indiana University Press, 1996). See also Herspring, "Samuel Huntington and the Study of Communist Civil-Military Relations," *Armed Forces and Society* 25, no. 4 (Summer 1999).

5. The classic work in this area is Roman Kolkowicz, *The Soviet Military and the Communist Party* (Princeton: Princeton University Press, 1967).

6. William E. Odom, "The Party-Military Connection, a Critique," in Dale Herspring and Ivan Volgyes, *Civil-Military Relations in Communist Systems* (Boulder, Colo.: Westview, 1978).

7. Timothy J. Colton, *Commissars, Commanders and Civilian Authority* (Cambridge, Mass.: Harvard University Press, 1979).

8. Dale R. Herspring, *The Soviet High Command, 1967–1989* (Princeton: Princeton University Press, 1990).

9. Adrian Keith Goldsworthy, *The Roman Army at War, 100 B.C.–A.D. 200* (Oxford: Clarendon, 1996), 251.

10. As cited in Goldsworthy, *The Roman Army at War, 100 B.C.–A.D. 200*, 149. See also "Roman Army Religion," in John Helgeland, Robert Daly, and J. Patout Burns, *Christians and the Military: The Early Experience* (Philadelphia: Fortress, 1985), 48–55.

11. As one author put it, "They were carrying the ark of Jehovah when the armed men were ordered to pass on before it. They were the custodians of the divine relics—the tables of the law, and Aaron's rod that budded." Henry Bradford Washburn, "The Army Chaplain," in *Papers of the American Society of Church History*, Second Series, vol. 7 (1923), 3.

12. As Hugh Elton put it with regard to chaplains in the Roman army, "Regimental chaplains are first heard of in the mid-fifth century in the East, though they may have appeared earlier, with Sozomon suggesting they were part of every regiment from the time of Constantine onwards." Hugh Elton, *Warfare in Roman Europe, AD 350–425* (Oxford: Clarendon, 1966), 90.

13. As quoted in Gordon Taylor, *The Sea Chaplains* (Oxford: Oxford University Press, 1978), 3.

14. See Jonathan Riley-Smith, *The First Crusaders, 1095–1131* (Cambridge: Cambridge University Press, 1997), 86–87, 99; Donald Queller and Thomas F. Madden, *The Fourth Crusade, The Conquest of Constantiople* (Philadelphia: University of Pennsylvania Press, 1997).

15. Washburn, *The Army Chaplain*, 7. Records also indicate that Cortez had chaplains. Indeed, both Portuguese and Spanish armies had chaplains in their conquest of the New World. "It may be said that chaplains accompanied all expeditions, for there was a universal conviction that the Church should follow its children withersoever they might go, and that the new lands should be conquered only in its name." Ibid., 10.

16. Washburn, *The Army Chaplain*, 11.

17. Throughout this study, I will refer to an army. Except when otherwise noted, I am referring to all of the country's services—not just those who wear an army uniform.

18. My discussion with Colonel Hans-Werner Weber a.D., November 1998.

19. Robert G. Hutcheson, Jr., *The Churches and the Chaplaincy* (Atlanta: John Knox, 1973), 17.

20. Hutcheson, *The Churches and the Chaplaincy*, 19.

21. Full citations for these sources can be found in the relevant chapter.

Part I
The Framework for Comparison

1

Chaplains in the American Military

The cross and flag are embodiments of our ideals and teach us not only how to live but how to die.

—General Douglas McArthur

Provided a ship has a good Commander and a good Chaplain you will never find anything much wrong with her.

—Admiral Sir John de Robeck

MORE THAN ANYTHING ELSE, a study of the history of the American chaplain shows conclusively that despite the many changes that took place in structure, uniforms, tasks, and rank, his central focus throughout the first 225 years of U.S. history was on motivation, morale, and political socialization, in addition to his spiritual duties. In a given period, one function may have been more important than in other periods—and the chaplain may occasionally have had to take on additional roles, such as schoolmaster—but, bottom line, he was indispensable when it came to fielding an effective military in times of peace and war. Even in the short period during the 1800s when the army was without chaplains, military leaders, not to mention those in Congress, soon realized that chaplains were vital.

Unlike the political commissars described in this book, the American chaplain never exerted control over line officers. In fact, the story of the American chaplain tells of his efforts to gain the autonomy and role definition necessary to carry out his job as a spiritual adviser, motivator, and morale booster. On some occasions, the chaplain himself did not know what his real role was. Historically, the military tended to stick him with tasks no one else wanted; only after 1920 was a serious and systematic effort made to determine where he fit into the U.S. military.

In the end, there were two primary justifications for the chaplain in the U.S. military. First, he provided the kind of spiritual help that most Americans believed was the constitutional right of the soldier, sailor, and Marine. Since the serviceman (and later the servicewoman) was not at home, he or she was unable to get the kind of spiritual assistance needed. It was the military's obligation to provide a clergyman—even if that individual was from a different domination or creed. Second, the chaplain played a crucial military role by helping to convince GIs that they were fighting a just fight and by working to motivate them and raise their morale. In the latter sense, chaplains were indispensable, especially during wartime. Indeed, one could argue that if the U.S. military had not had chaplains, it would have had to invent them. Who else would worry about things like the individual soldier's or sailor's personal problems or intercede with the appropriate officials when disciplinary violations occurred? Line officers, including commanders, were concerned about anything impinging on morale, but they did not have time to get involved in such matters—except in emergency situations. Furthermore, when it came to combat, someone was needed not only to deal with the wounded and dying but to provide young Americans with the kind of motivation and personal help they needed as they were called upon to put their lives on the line. These duties clearly went beyond the line officer's job description. As the history of the U.S. chaplaincy shows, on many occasions the role played by such men was critical to the outcome of the battle. The case of navy chaplain Joseph O'Callahan, S.J., discussed further on, is only one example.

Although chaplains played important roles in the military, we must keep in mind that their position evolved over the years. The chaplains of 1997, 1943, 1914, 1864, and 1775 might all have prayed with a soldier—indeed, they might have used the same prayer—but their relationship to the rest of the military would have been quite different. It is to this evolving relationship that we now turn.

The Continental Period

The early Americans followed the English example—much like that which was practiced at the time of the English Civil War. In the latter case, the lord of the manor supplied not only the equipment needed by most of his troops, he also had a private chaplain who dutifully followed his lord into battle when the time came. It was the chaplain's task to provide the officers and men with the spiritual guidance and comfort needed to keep their morale up, as well as to motivate them on the eve of a major battle. The major difference in the American case was that the English aristocratic system was not nearly as strong in the New World as it was in the Old. As a consequence, a militia's or military unit's chaplain might well have been the local pastor. He might have been elected, appointed, or just available as the militia went into battle.

The first recorded case of a chaplain being involved with American troops occurred in the spring of 1637. A group of settlers put together ninety men in response to a raid by Indians. The unit was led by Captain Mason, who took along the local pastor, Samuel Strong, as chaplain.[1] As far as the chaplain's duties were concerned, Captain Mason faced the question of whether or not to continue following the Indians into the forest—an action most of his officers opposed. Mason turned the issue over to Chaplain Strong, who prayed all night and the next morning reported that the unit should proceed to attack—which it did!

In addition to supporting the army or militia in battle, continental chaplains accompanied small scouting parties or found themselves stationed at isolated posts during periods of peace. For example, on June 27, 1702, Massachusetts appropriated twenty pounds toward the support of a chaplain at Brookfield—a town that was fortified and garrisoned after the preexisting town had been wiped out by Indians.

Indeed, the continental commanders expected that they would be supplied with chaplains. For example, in 1725 Captain Henry Dwight, who was in command of a fort in what is now Vermont, wrote, "We shall lead a heathenish life unless a chaplain be allowed." As a consequence, a man by the name of Daniel Dwight was appointed and served for two years with a salary of 100 pounds in that position. By 1756, the monthly pay of chaplains was set at 6 pounds 8 shillings, "which was one pound more than the pay of a captain and half the pay of a colonel."[2] At this point, the morale issues the chaplain was expected to take care of in the American military included things such as discipline, respect for orders, avoidance of drunkenness, clean living, and so on.

The French and Indian War, which lasted from 1755 to 1763, had little impact on chaplains. By and large, they were assigned to particular regiments. In some instances, however, they were stationed at forts. Normally, they were appointed by the governor of the province in which the troops were raised or in which the fort was located. A total of thirty-one chaplains served colonial troops during that period. Nearly half came from Massachusetts. Most were Congregationalists, although a large number of Presbyterians and some Episcopalians were also appointed.[3] By and large, the chaplains had the same religious background as the men they served. Usually, most, if not all, of a town's or region's inhabitants were of the same religious background. The pastor would be "their" chaplain, so the question of religious orientation or the need for the chaplain to serve those of other faiths was not a serious concern.

The Revolutionary War

The War for Independence presented the country with a new and different problem. No one doubted the importance of chaplains—indeed, that issue was not even questioned. To cite only one example, George Washington had long been convinced of the importance of chaplains and the role of religion in the military. During the French and Indian War, he made it clear that one of his priorities was to find a "gentleman of sober, serious, and religious deportment, who would improve morale and discourage gambling, swearing and drunkenness."[4] From Washington's perspective, chaplains were important not only because of their religious role, but because they contributed positively to the military efficiency of the unit. As Williams put it, "Behind his attitude lay in part a British conception of a chaplain as a brother officer and gentleman on the staff for . . . the soldiers but also in part the squire's idea of a divine who would assist in professionalizing 'the commonalty of a colonial militia.'"[5] Or as another author put it,

> As before hostilities commenced, there was scarcely a military muster at which they were not present, exhorting the militia to stand up manfully for the cause of God— on some occasions saying, "Behold God himself is with us for our captain, and his priests with sounding trumpets to cry the alarm"—it was to be expected, when war actually broke out, they would be found in the ranks of the rebels, urging forward what they had long proclaimed as a religious duty.[6]

In other words, for leaders like Washington, chaplains would help unify the military by building a common weltanschauung, while at the same time battling drunkenness and gambling. Equally important—and here we see a similarity between the American chaplain and what would become the Soviet and East German political officers—he would also be expected to help maintain a high level of troop morale.

In the actual selection process, not only did the man's religion play an important role, but political influence was crucial. In many cases his knowledge of Indian languages, French, and German was also important. The chaplain's term of service ranged from a few days to eight years, and chaplains were often called upon to serve not only as surgeons but in some cases also as commanders. They were always subordinate to the senior unit commander, however.[7]

The chaplain was helped by the fact that soldiers were often required to attend church services. For example, the Massachusetts Anticipatory Regulations of early April 1775 ordered officers and soldiers not only to go to church services but to "deport themselves properly, the former on pain of court-martial and reprimand, the latter of a monetary fine."[8] By and large, Congress adopted the same regulations in Articles of War on June 30, 1775. This provision was retained when the articles were revised on September 20, 1776. Irreverence was not acceptable behavior for either officers or men.

Regarding their pay, regimental chaplains received $33.50 per month. Their terms of service varied. Like surgeons—with whom they would be compared over the years—chaplains signed contracts for periods of six to twelve months. Chaplains at other levels were paid $20 per month—an amount that placed them on a par with army captains. Indeed, this would become a major source of debate over the years—how to rank and pay chaplains. At this point, they did not have a rank but were treated as officers. However, they were not commissioned officers in the strict sense of the term, and their pay was based on their level of responsibility (e.g., a regimental chaplain received more than a company-level chaplain).[9]

The vast majority of chaplains did not take part in combat and their noncombatant status was recognized by both sides—although there were exceptions.[10] In addition to providing regular religious services and counseling, their primary purpose was to raise the morale of the troops on the eve of a battle by instilling in soldiers a sense of purpose; chaplains were also expected to be present in the heat of battle to help motivate those who wavered. In a few instances, chaplains actually assumed command, but these were clearly exceptions. They were also used as intelligence officers, with the task of interrogating prisoners. Because chaplains tended to be educated and were members of the clergy, it was felt that many prisoners would speak with them more openly than with regular officers.

As is often the case, the navy operated differently. It was up to the ship's captain to find a chaplain. There was no regular procedure for recruiting or hiring them. In some instances, captains hired unordained men who also worked as surgeons or clerks—in addition to conducting divine services. Their status was similar to their role in the army, in that they were considered junior officers and were given an appropriate amount of any prize money their ship received. Smaller vessels often rated only a schoolmaster. As was the case in the British Navy and the French Navy, however, this was generally a distinction without meaning, because chaplains were almost always called upon to work as schoolmasters at sea. There were no naval academies, and someone had to take on the task of teaching the midshipmen how to read, write, and do the mathematical calculations necessary for them to function as officers. As in the British Navy, they also were often used as secretaries to the commanding officers.[11]

Were chaplains valuable during the Revolutionary War? Evidence suggests that the answer is yes. As one writer put it, "The influence in support of discipline and morale was so valuable that chaplains have been considered an essential part of the military organization in all emergencies."[12]

One key problem that would face the newly created American military was the issue of religious pluralism. During the Revolutionary War, the problem had been relatively simple. The religion of the chaplain and that of the unit in which he served were almost always the same. However, that was a unique situation, and it reflected the practice in the colonies and Europe of creating an established church. With the exception of Rhode Island, all of the colonies had established churches—and chaplains were generally appointed from the prevailing religious institution. Many chaplains were appointed by the governors of the states—and the majority came from New Hampshire, Massachusetts, and Connecticut. As a result, most chaplains were Congregationalists since that was the predominant religious orientation. In all, 72 were Congregationalists, 32 were Presbyterians, 18 were Episcopalians, and 9 were Baptists.[13] In other instances, however, the officers themselves made the selection—as in Rhode Island and Virginia. In time, the same thing happened with units from New Hampshire. There were even instances when the chaplain was appointed by the state legislature (for example, from New York and Pennsylvania).[14] In the navy, most chaplains were Episcopalian—a situation that would not change for many years, largely as a result of the strong influence the British Navy would exert on the creation and evolution of the U.S. Navy. Indeed, by 1840 49 percent of all navy chaplains were still Episcopalians.[15]

The Postrevolutionary Period

As often happened in American military history, once the war was over, the size of the military shrunk radically. In May 1783, for example, Washington recommended that the army be reduced to only 2,631 officers and men. This meant the creation of 4 infantry regiments—each having 477 officers and men. The remaining 723 were to be formed into a regiment of artillery. Each regiment would have its own chaplain—thus giving the army a total of 5 chaplains.[16] The chaplain received the same pay as the unit's surgeon—$40 per month, one food ration, and enough forage for one horse.[17] This placed chaplains somewhere between a lieutenant and a captain of infantry.

As far as the functions to be carried out by these chaplains was concerned, political socialization figured prominently, as Secretary of War Henry Knox observed,

> Every legion must have a chaplain, of respectable talents and character, who besides his religious functions, should impress on the minds of the youth, at stated periods, in concise discourses, the eminent advantages of free governments to the happiness of society, and that such governments can only be supported by the knowledge, spirit, and virtuous conduct of the youth—to be illustrated by the most conspicuous examples in history.[18]

What is important about this quote is that it illustrates very clearly that in addition to the kind of morale building and religious leadership that would be evident in Cromwell's army, the American chaplain was expected to play a role in nation building—in convincing the average soldier of the rightness, even righteousness, of the government he served.

Meanwhile, the U.S. Navy retained only one chaplain on active duty. He was an ordained minister but seems to have been kept on primarily because of his ability to teach mathematics. In fact, this individual later created a one-man school for midshipmen and formally requested that his title be changed from chaplain to naval mathematician.

In 1814 Congress authorized both a chaplain and a schoolmaster for the seventy-four new gun ships of the line, thus relieving the chaplain of the need to function as both schoolmaster and spiritual leader. However, the regulations adopted four years later made it clear that in addition to his religious duties, the chaplain was to serve as the commodore's secretary when called upon. In essence, the chaplain's job was almost completely undefined—beyond holding religious services. An individual could shift from being a chaplain to a clerk, to a purser—and back. In the eyes of many commanding officers, the chaplain's religious duties took second place to his other, more important jobs—except when it came to baptizing babies—a situation brought about by the presence of wives aboard ships. In some instances, ships' captains appointed a member of the crew as chaplain, thereby making it clear that specialized religious training and ordination were not prime prerequisites for the job.

By 1842 the number of navy chaplains had increased to twenty-four, partly a result of the navy's decision to increase their pay to $1,200 per year. Further pay increases followed in 1852 and again in 1860, until chaplains were paid the equivalent of a full lieutenant. While it is hard to determine with any exactness just how influential chaplains were in the navy during the first half of the 1800s, they did play an important role in curtailing flogging, in helping create shore schools such as the U.S. Naval Academy, and in getting the navy to pay sailors in lieu of their grog ration.[19]

The situation for chaplains in the army was different. As war appeared on the horizon in 1812, Congress ordered the immediate creation of 13 regiments. These units were to be large—numbering more than 2,000 officers and men in an infantry unit and almost that many in an artillery outfit. Congress was more stingy when it came to chaplains, however. They were authorized, but only at the brigade level. In addition, militia units were called to the colors—but many of them came with their own chaplains. All in all, based on the records available, the full allotment of chaplains does not appear to have been utilized during that war. When the war was over, only four regular army chaplains were still on active duty. The others were probably civilian clergymen who were kept on contract. There were no Jewish chaplains, and, as Williams noted, "While there were undoubtedly priests on the battlefield in New Orleans, there is no clear record of an official Catholic chaplain."[20] The brigade chaplain was paid as and given the same status as a major of infantry.

After the War of 1812 the army again kept only one chaplain on active duty. He was assigned to West Point, where he served as professor of geography, history, and ethics—with the same pay as a professor of mathematics. Regarding the rest of the army, primary reliance continued to be placed on the contract system whereby a civilian clergyman would agree to serve as a military chaplain for a specific period of time. In addition, in more remote regions, Indian missionaries and itinerant preachers were used. Their salary was paid by voluntary contributions, not by the U.S. government.

Upset at what he considered the low moral standards in the army, in 1831 Secretary of War Lewis Cass argued that a chaplain paid by the U.S. government should be stationed at every army post. Despite Cass's expression of concern, nothing was done until 1838 when the post chaplain system was established.[21] This act called for the post to employ a chaplain who should "also perform the duties of school master." The total number of these chaplaincies was limited to twenty and could be set up only in those places "most destitute for instruction." While these posts were intended for the more sparsely populated West, most of them were set up in the East and, more often than not, the occupants were not ordained ministers but senior sergeants or cooks who were given the salary and allowances of a captain of cavalry, rations for four people, and quarters on the post. In 1841 church attendance was made mandatory throughout the army.[22]

The Mexican War was important for the development of the American chaplaincy because it marked the first time that Roman Catholics could be chaplains. President Polk had two major concerns as the United States entered that conflict. First, a lot of Irish recruits had enlisted in the army, and the president feared that if he did not appoint a few Catholic chaplains, some of these troops might desert—especially since the United States was fighting against a largely Roman Catholic country. Second, Polk wanted to make a gesture that would show the Mexicans that the United States was not anti-Catholic and would not attempt to force its religion on that country. What better way than by appointing two Catholic chaplains!

With the end of the Mexican War the army went back to the system of post chaplains, although the number of chaplain slots was increased from twenty to thirty. At least four Catholics received regular appointments as chaplains. "Two held successive positions at the post at Monterey. One served at Fort Belknap and another was selected by the council of administration at Camp Floyd in 1859."[23] All in all, between 1813 and 1856 only eighty chaplains received appointments in the army.[24]

In the meantime, the military chaplaincy came under considerable criticism from outside—primarily from those who maintained that its existence was a violation of the principle of separation of church and state. In the end, nothing came of the controversy, but it did signal a problem that would come up again and again, as many would claim that taxpayer support for the chaplaincy—including congressional chaplains—violated the establishment clause of the First Amendment. Chaplains' actual responsibilities, in addition to conducting religious services, included educating soldiers as well as the children at the base.[25] Indeed, chaplains were assigned to a post—not to the troops. As a result, there were no chaplains to accompany soldiers when they went to the field. For example, in his autobiography, Percival Lowe noted two occasions when he dealt with chaplains, and in both instances they were located on bases such as Fort Leavenworth or Fort Riley.[26] The idea that chaplains spent most of their time at a post would change around the turn of the century, but during the nineteenth century, chaplains were only available when the soldier returned to his base—assuming, of course, that the post had a chaplain.

While it may be considered a small, relatively unimportant matter by civilians, the issues of uniforms and rank were matters of great concern in the armed forces. On the one hand, there were those who believed that chaplains should be set apart from others in the military—both officers and men. They should be treated with respect, but since they were mostly noncombatants, they should not be commissioned officers. Another concern—which continues to be argued today—was that if chaplains became full-fledged officers, their task of relating to the troops would be more difficult. The military is a hierarchical organization. Officers are officers and enlisted men are enlisted. Even noncommissioned officers, who serve as a bridge between the two, are not really officers. How to circle the square? How to make chaplains both accessible to the troops and yet ensure that they had the authority they needed to carry out their tasks? A variety of approaches were tried—all of which led to changes and modifications in uniforms and insignia over the years.

Navy chaplains originally wore only clerical attire. In 1830 this meant "plain black coat, vest and pantaloons" or "black breeches." As noncombatants, they were forbidden to carry swords. By 1838 they were permitted to wear navy brass buttons with eagles on their "uniform." On the frontier, it was common for army chaplains to wear whatever clerical outfit was most appropriate for the denomination they represented. In the regular army, chaplains were ordered to wear the blue single-breasted coat with ten gilt buttons of the general staff, but without the gold epaulets, indicative of rank. By 1821 they had been strictly forbidden to wear any uniform at all. Then they were told to wear black "citizen's dress with buttons of the corps of Engineers, round hat, black cockade with gold eagle." By 1832 they were permitted to wear a dress sword, only to learn in 1840 that they were back to no uniform.[27]

Lest the reader get the wrong impression, this uniform game was not a game at all. It went to the heart of what a chaplain was supposed to be. Like commissars and political officers, the chaplain was the only member of the military who served two gods—the army and his religious denomination. Many—including a number of chaplains—believed that if they were seen as just another officer, their uniqueness would be undermined. Besides, a number of senior officers did not want chaplains around under any circumstances—some of them believed that they cost more than they were worth. How could one impose the kind of iron discipline so necessary to military life if a chaplain was looking over the commander's shoulder all of the time, criticizing him whenever he "got tough" with the troops?

Another concern on the part of many denominations was that if the chaplain became a full-fledged officer, they would lose control over him. He was treated as if he was an officer—wasn't that enough? The denominations did not want the military brass getting between them and their clergymen.

There was another side to the story, however. Without an officer's rank, the chaplain was at the mercy of any commanding officer who might be antireligious or anticlerical. The commanding officer controlled everything on post—from the schedules for the troops to whatever kinds of resources might be available. He could even control the chaplain's life and time by assigning him unpleasant tasks—or jobs that took away from his primary duty of dealing with and helping soldiers. Obviously, if the chaplain carried an officer's commission, he would have a commensurate amount of authority and be able to deal with other officers on a more even playing field. The issue of whether a chaplain was or wasn't an officer remained unresolved as the Civil War approached.

The Civil War

The Confederacy did not make a serious effort to create a functioning chaplaincy. To be sure, the Confederate Congress authorized Jefferson Davis to assign chaplains to regiments. However, only a few months later, this same Congress cut the chaplain's pay. The Congress's reasoning was that the chaplaincy was a calling and not a way to earn money. Many believed that cutting money would ensure that only the most idealistic became chaplains. In the end, it had the opposite impact. "By the spring of 1862, more than half of the Southern regiments lacked military clergy."[28]

Unlike their fellow clergymen in the North, chaplains in the South spent much less time on morale and discipline-related issues. The evangelical Southern chaplain saw his primary task as "saving souls." These men were going into battle and their eternal souls were on the line. Victory for the South was important, but helping soldiers gain salvation was even more important.

The situation in the more structured and better organized Northern chaplaincy was quite different. To begin with, on May 3, 1861, General Order Number 15 was issued. This order "provided for the organization of volunteer regiments, including the appointment of a chaplain chosen by the vote of the field officers and company commanders."[29] General Order Number 16 provided for the same provision for the regular army. Technically, regimental commanders who did not have chaplains were permitted to appoint them. In practice, however, the commander often delegated this responsibility to the unit's officers, who would interview a number of candidates and then select the one they considered most appropriate. In some instances, a minister would enlist as a regular soldier and then apply for the job. Sometimes it worked—the officers would come away with a feeling that they had picked one of their own. The only requirements that the individual had to satisfy were that he had to be an ordained minister of some Christian denomination and that he had to be approved by the state governor.[30] The main downside to the election of chaplains was that overall quality was hurt. The most popular individual in a unit was not always the best qualified to take on this difficult job.

There were also chaplains who were anything but paradigms of morality—especially during the early stages of the war, when many unqualified men entered the chaplaincy. For example, there are reports of chaplains who sold items to their troops that had been donated by benevolent societies, another story of a chaplain who lived in a brothel while his troops were in the field, and still others about chaplains who ran at the first sound of gunfire.[31] In short, "On the eve of the Civil War . . . the military chaplaincy was probably at its lowest point, both militarily and ecclesiastically."[32] The situation changed significantly, however, after Congress passed legislation in 1862 that required chaplains to be ordained ministers in good standing with a recognized religious denomination. In addition, commanders were to evaluate chaplains within thirty days and to dismiss those deemed unfit.[33]

The tasks assigned to a chaplain varied. First, and foremost, it was his responsibility to organize and conduct religious services, to bury the dead, and to inform the next of kin of those who died. In addition, he was expected help maintain discipline and morale and to engage in political socialization by convincing soldiers of the rightness of the North's cause. To quote one Civil War chaplain, "Now, comrades, we have got orders to march, and I must stop. God bless you, and make you faithful soldiers for God and your country."[34] The chaplain's efforts in this area appear to have been effective. As one officer remarked, "We count our chaplain as good as a hundred men in a fight, because the men fight so much better when he's with them."[35] H. Clay Trumbull's sermons also blended the theme of patriotic duty with "the evil nature of the Confederate rebellion and the ultimate victory of the Union forces."[36]

Chaplains also became directly involved in combat—in both armies. Take, for example, the case of the Northern chaplain Father William Corby:

At Antietam, Corby galloped in front of the Irish Brigade as they were rushing into battle, told the men to make an act of contrition, and gave a last absolution. Within a few minutes, 506 of them had fallen. At Gettysburg he gave general absolution to Hancock's corps a moment before they joined in deadly strife near the spot where his statue now stands.[37]

The same was true of the South:

I belonged to Company G. 6th Texas Cavalry Regiment, Ross's Brigade. We were dismounted and served as infantry in the battle of Cornith, and our regimental chaplain was Verderhurst, a talented young minister from Waco, Tex. As we were about to assault the strong works of the enemy on the morning of the 4th of October, 1862, he came to our company with a gun, went into the charge with us, and in the awful slaughter that followed was shot dead.[38]

While few soldiers would intentionally shoot at a chaplain on either side, the fact was that many chaplains placed themselves in harm's way in an effort to provide the kind of religious and moral support needed by their troops.

This kind of behavior occurred in spite of the special status enjoyed by chaplains during the Civil War. A general order was issued by Washington that stated that chaplains should not be held as prisoners of war. According to this order, any chaplains held by the North were to be immediately released. The South issued a similar order.[39] In spite of this assumption of noncombatant status, chaplains were constantly in the front lines. Indeed, those who were the most evident in battle were also the most effective. As one writer remarked, "Those chaplains who could combine spiritual ministrations with skill in bandaging wounds and relieving suffering gained an especially large place in the affections of the men."[40] To quote another soldier, "Without a thought of his personal safety he was on the firing line assisting the wounded, praying with the dying, doing all that his great loving heart led him to do. No wonder our boys love our gallant chaplain." Or as an officer noted, "I am particularly proud and thankful for him as some officers (nonprofessors) used to think and even say that a chaplain was a sort of fifth wheel . . . and even voted against having one, but now all are ready to admit that we could not get along without our chaplain."[41]

Chaplains could be and often were assigned tasks that went far beyond the spiritual, such as paymaster and postmaster, as well as the traditional job of counseling soldiers about drinking and gambling. Then there was always the task of listening to soldiers' troubles. During the Civil War, some 2,300 men served as chaplains in the Union Army in a variety of functions. For example, of the 1,068 on active duty in 1863, 21 served as post chaplains, 117 as hospital chaplains, and 930 as regimental chaplains.[42] Chaplains also helped soldiers write letters home. To quote Father Corby, "Hundreds of such letters passed homeward, and in time the dear ones would write to the chaplains of the brigade, asking for more information."[43]

As in the past, chaplains also played an important educational role. Many soldiers were not only illiterate, but a large number of them spoke only rudimentary English. Chaplains organized courses in what we would today call "English as a foreign language," while others set up classes in debating, Latin, German, mathematics—whatever the resources on hand and the needs of the soldiers dictated. As one chaplain put it,

> Those men who attend regularly when off from duty are making good proficiency in the branches taught, viz,: Reading, Writing, Spelling, Geography: Mental and Written Arithmetic. Men who do not attend the school are taking lessons in their quarters, of those of their comrades who are more proficient.[44]

The Civil War also marks the first time a Jewish chaplain was appointed. The federal statute of 17 July 1862 dropped the proviso that a chaplain must be a Christian. As a result, the Fifth Pennsylvania Cavalry elected Michael Allen, a Jewish soldier, to be regimental chaplain. While he did a good job, he served at a time when there were a lot of complaints about unqualified chaplains. Allen did not meet these qualifications and, as a consequence, he was told to resign his chaplaincy and did so.

In one sense, the Civil War marks a major turning point in the development of the chaplain's role in the American military. During the revolutionary period, there was only a limited need for chaplains to accommodate the needs of individuals from other religions. After all, the units they were assigned to tended to be homogenous from a religious standpoint. This was less true of the Civil War—although the situation was a far cry from the military of the 1990s. It was gradually becoming obvious to all concerned that in the preparation of men for battle, not to mention ministering to them in combat, one could not stand on theological fine points. If the soldier was Protestant and the chaplain Jewish, it was up to the chaplain to minister to the young man in a way that was meaningful to him. Similarly, a Baptist chaplain dealing with a dying Catholic might well find himself saying the Rosary. As Armstrong noted, "To be narrow minded, theologically or in regard to conduct, drastically limited the chaplain's usefulness."[45] In addition to becoming "all things to all men," the chaplains were also learning a very important lesson, one that civilian society would take many more years to learn: that the only way to get anything done was for all of the chaplains to work together. Indeed, one constant theme that runs through chaplains' writings after the Civil War was how much they were dependent on the help and assistance of members of other religious orientations. This did not mean that by helping the Catholic, the Baptist was untrue to his own religious convictions or that the Jewish chaplain was not true to his religious orientation if he were to listen to the young man's testimony of faith in Jesus Christ or even help him make such a testimony. The chaplain corps was beginning to learn that in combat, one must rise above one's own denominational orientation and place one's parishioner—in this case, the soldier or sailor—first. Father Corby probably put it best when he observed that "One good result of the Civil War was the removing of a great amount of prejudice. When men stand in common danger, a fraternal feeling springs up between them and generates a Christian, charitable sentiment that often leads to most excellent results."[46] Occasional complaints of discrimination in favor of one orientation or another would still arise, but the U.S. military had learned an important lesson about the role of chaplains and religious pluralism during the Civil War.

In a nutshell, the Civil War taught chaplains (and the military) two things: the need for religious tolerance and flexibility and that chaplains were indispensable.

Post–Civil War Period

Navy regulations of 1865 were clearer than they had been in the past in outlining the chaplain's duties. In addition to conducting religious services, he was to continue to function as a schoolmaster by providing both Christian and elementary instruction for the sailors. He was also to visit and comfort those who were sick. Indeed, the Navy regulations of 1893 stated that the chaplain's primary duty during combat was to "aid the wounded."[47] Finally, he was to prepare a quarterly report, as well as one for each cruise, on how he had carried out his duties. Based on what we know of the activities of navy chaplains, it appears that a chaplain was assigned to each squadron and the commander had no say in the matter. Many did not want chaplains, fearing that they would undermine the squadrons' rigid concept of discipline, but from this point onward commanders would have to live with them.

With the end of the Civil War, the number of navy chaplains was set at twenty-four. What was most surprising was that as the size of the navy was expanded toward the end of the century and the beginning of the next, the navy held the number at twenty-four. Only in June 1914, as the First World War was about to begin, did Congress raise the number of chaplains authorized to serve in the navy to forty—a number that assumed there would be one chaplain for every 1,250 naval personnel.[48]

An improvement was made in the status of navy chaplains in 1906 when Congress changed the pay system. They were now commissioned officers with the rank and remuneration of a lieutenant (jg).[49] By 1916 the pay that chaplains received was on a par with that received by other naval officers. Indeed, for the first time in the navy, chaplains had achieved full equality with other officers.

After the Civil War, the size of the army's chaplain corps was also reduced—to only thirty post chaplains. They were not commissioned officers but were employed by the post's council of administration—again, on contract. They were assigned the duties of a chaplain and a schoolmaster at those posts that were considered the most isolated and most in need of instruction—once again highlighting how important the chaplain's educational function was to the army. In 1866 Congress authorized six more chaplains—for each of the six black regiments in the army. These men were called regimental chaplains, and they were commissioned, although they held the rank of "chaplain without command." They were paid the same as a first lieutenant and were expected to oversee the "instruction of enlisted men in the common English branches of education."[50] Unfortunately, there were almost never enough chaplains—black or white—to meet the educational needs of the army's far-flung units.[51]

A year later, Congress authorized commissioning post chaplains "now in service or hereafter to be appointed" and made them the equivalent of a "captain of infantry." This made all chaplains in the army equal with regard to things like terms of office, retirement, allowances for service, and pensions.[52]

The requirements to become a chaplain had not changed significantly from what they had been prior to the Civil War. The applicant had to be ordained by a recognized denomination and had to be recommended by an ecclesiastical body or five ministers from his denomination. In their religious orientation, most were Episcopal. The first Catholic priest to be made a regular member of the chaplain corps was not appointed until 1872 (with seven more to follow by the outbreak of the Spanish-American War), and the first Jewish rabbi would not become a chaplain until World War I—and then as a reserve, not a regular, officer.[53]

One problem that haunted the army's chaplains during the second half of the nineteenth century was the tendency of commanding officers to dump jobs on them that no one else wanted to do. In the 1870s, for example, in addition to his normal religious duties, the army chaplain worked as a schoolmaster, librarian, post gardener, post treasurer, and manager of the post bakery. Indeed, the chaplain had so many different jobs that he was often referred to as a "handyman." In spite of this expansion of the tasks he was expected to perform, the chaplain continued to have responsibility for morale, motivation, and political socialization. The problem, however, was that he was being asked to do so many things that his ability to carry out these tasks was threatened. Sometimes chaplains had an enlisted man assigned to them and that helped. However, it did not solve the problem.

In addition to illiterate black soldiers, a number of immigrants in the army barely spoke English. Just teaching them to read and write the English language was a major challenge. As a result, the chaplain worked long hours on projects that had little to do with his religious duties. This not only sapped his energy, it also went contrary to what many chaplains believed was their primary function. The situation improved somewhat in 1878 when the War Department assigned instructors to visit the various posts and oversee the educational process. Then in 1894 Congress passed a law stating that individuals who could not speak English at a thirteen-year-old level could not enlist in the army, thereby eliminating the need to teach English as a foreign language. Finally, in 1895 army regulations authorized post commanders to assign officers—other than chaplains—to supervise the post school. While some chaplains continued to oversee education at posts, it was no longer part and parcel of their job description and thereby marked an important step forward.

This is not to suggest that problems did not exist. To begin with, many chaplains were political appointees—individuals who were unqualified and unworthy. They also did not fit into the hierarchy of social customs in the army. A lot of regular officers did not like them and saw no reason for their presence. Besides, chaplains were commissioned at the rank of captain, at a time when it took regular officers many years to reach that level, and chaplains occupied often scarce officer quarters. Most of all, however, a lot of officers were upset at the constant battle the chaplains waged against drinking and drunkenness.

An effort to reform the chaplaincy was undertaken when a bill was introduced in Congress in 1890. In essence, this bill provided for more structure among chaplains. It added 15 chaplains to the 34 already authorized. The difference was that the latter came in a more junior position—as the equivalent of a first lieutenant—rather than in the rank of captain. In addition to his other duties, the chaplain was expected to supply soldiers with both secular and religious literature. Chaplains also supervised libraries, reading rooms, and gymnasiums. The bill called for a more rigorous selection process and urged the creation of a professional development program for chaplains. Most important for the future of chaplains, however, was the call for them to report on the causes of discontent on the part of enlisted men. The latter idea particularly incensed regular officers, who saw it as a challenge to their authority. Unfortunately, the bill was never enacted. Chaos would continue to reign in the ranks of the army's chaplains for a few more years.

But the push for reform of the army's chaplain corps would not go away. To begin with, no central authority supervised chaplains, someone to whom chaplains could turn when problems appeared. In addition, more often than not the decision on whether a chaplain had what he needed to do his job was determined by his local commander. If that officer had other priorities or believed that chaplains were unnecessary, the chaplain would find it very difficult to do his job. As one author put it,

> Success depended in great part on the sympathy and support of the post commander. Since many if not most officers were contemptuous or suspicious of them, many chaplains were defeated before the battle even began.[54]

To make matters worse, there was no educational program to help chaplains grow. Furthermore, since they could not be promoted beyond the rank of captain, nothing existed in the way of an incentive to encourage them to do better. They had the opportunity to make suggestions through military channels, but commanders looked down upon chaplains who criticized the way things were run. In short, when it came to policy matters that had a direct impact on how the chaplaincy functioned, chaplains had little or no say.

An important step was taken in 1899 when the upper age requirement for an applicant was lowered to forty-four, and he was required to pass an examination "as to his moral, mental, and physical qualifications as may be prescribed by the President."[55] In fact, the examination that chaplains had to pass was the same as the one given to those seeking a regular commission. It covered fourteen subjects, such as arithmetic, algebra, geometry, trigonometry, surveying, logarithms, grammar, writing from oral dictation, English literature, geography, orthography, international law, the U.S. Constitution, and history. The board examining the applicant had the authority to overlook a low score in one area as long as scores in the others were sufficiently high.

Regulations were further tightened up two years later when the maximum upper age for an applicant was reduced to 40. The number of chaplain billets was increased from 34 to 57. The new rules also stipulated that the applicant had to be an American citizen and that a medical officer had to be represented on the board of examiners.[56]

Rank again was addressed in the 1904 Congressional Act. It authorized the president to promote chaplains from the grade of captain to that of major—after the person had served for a total of ten years as a captain. This was the first time any chaplain could hope to rise to this rank. A structure was now emerging. All new chaplains would be commissioned as first lieutenants. They would be promoted to captain after seven years' service. Those not selected for promotion to major would be promoted to the grade of a captain of cavalry. Five years later, the army decided to assign junior chaplains to more senior ones so that they could "learn the ropes." Progress was being made, but it would still be some time before the chaplain corps would have the structure it needed to carry out its mission effectively.

World War I

The United States reacted to World War I in much the same way it had to wars in the past. Overnight, there was a major increase in the number of soldiers—with a requisite need for more chaplains. The Act of 12 May 1917 called for the assignment of one chaplain per regiment and one for each 1,200 soldiers in the coast artillery. The problem is that this act did not take into consideration the large number of service troops in the army or the fact that the regiments were much larger than in the past—up to 3,600 men in a unit. This was clearly far more than one chaplain could handle. In time, chaplains' numbers would increase.

When the United States entered World War I, a total of seventy-four chaplains was in the regular army. Seventy-two chaplains from the National Guard soon joined them, as did hundreds of civilian clergy. By the end of the war, there would be 2,217 chaplains in the army.[57] An important development during the First World War was the decision to set up a chaplains' school. The course lasted five weeks and included subjects such as military law, international law, army regulations, organization, insignia, customs, and military hygiene. Theology was a matter of private concern between the individual and his denomination. It was the army's task to train chaplains how to function in a military environment. Those considered unfit would be dropped from the school, while those who completed it successfully were commissioned as first lieutenants. Instruction began at Fort Monroe on March 1, 1918. A total of 1,042 completed the school, while another 273 entered the school but failed to graduate. Unfortunately, the school was deactivated during the middle of January 1919.[58]

The issue of insignia continued to be a major concern to chaplains. The first one (adopted in 1880) was the shepherd's crook because it symbolized their pastoral function. It was replaced by a plain Latin cross in 1888, which remained the chaplain's symbol until 1918, when Jewish chaplains entered the military and were allowed instead to wear a device composed of the tablets with a star of David over it. The debate over rank also continued during and after World War I. In 1914 chaplains were permitted to wear the uniform of a staff officer. Then in 1918, rank was removed because of the recurring concern that it would hinder chaplains' work with soldiers. While many believed that this action denigrated the role played by chaplains, the situation remained that way until 1925, when ranks were reintroduced.[59]

One of the more encouraging aspects of the chaplaincy during World War I was the close way in which all of the chaplains worked together. As they had learned in the Civil War, the battlefield was no place for sectarian differences. A dying soldier did not care what the chaplain's religious background was—all he wanted was to be comforted in the way he understood best, and it was up to the chaplain to make that happen.

One of the more unusual duties assigned to the chaplains during World War I was that of censoring mail. On occasion, this task used up half of the chaplain's day. To make matters worse, his relationship with his congregation was inevitably impaired if he found it necessary to cut out parts of their letters—if he believed the letters compromised national security. In short, this additional duty put the chaplain in a very difficult situation.[60]

When it came to the use of weapons and involvement in combat, the army's experience in the First World War was not much different from the way it had been during the Civil War. By and large, chaplains did not carry weapons and were expected to be treated as noncombatants. On the other hand, stories were told about chaplains who helped drive back the enemy in an emergency. For example, one chaplain was cited for "caring for the wounded, encouraging the advancing troops, and throwing hand grenades without thought of personal safety."[61] This situation was not unique to the American chaplaincy, however. The primary history of the British chaplaincy tells of a Father Gleason of the Munsters, who, during World War I, "when all of the officers were killed or wounded at 1st Yepes, had stripped off his black badges and taken command of survivors and held the line."[62]

The bottom line when it came to chaplains in World War I—from the army's standpoint—was that the chaplain was expected not only to provide for soldiers' spiritual needs but to help motivate them when the going got tough. This was evident in the decision to appoint Julian Yates, who would later become the army's chief of chaplains, to the War Plans Division to be assistant officer in charge of morale training in 1919. Only well-motivated soldiers, who were convinced of the rightness of their cause, could be expected to fight hard.

The Interwar Period

The lack of a well-organized structure to supervise chaplains remained a serious problem. In 1918 an effort was made to deal with this problem—but only in Europe. An executive committee was set up to study the conditions under which chaplains served and to make recommendations. In addition, this committee assigned chaplains to units and installations in Europe and supervised the Chaplain's School in France.[63] Unfortunately, these "organizational" efforts had little impact outside of Europe. Indeed, the United States' recognition that things had run much better in Europe was behind the introduction of meaningful reform into the U.S. chaplaincy.

The key problem was the command relationship. Chaplains did not command or control other chaplains. If the commander refused to provide the chaplain with the support he needed to carry out his mission, there was no place he could turn. His situation was the direct opposite of the commissars in France and the Soviet Union, where a commander ignored them at his peril. U.S. commanders could and did tend to ignore chaplains on occasion. And there were other problems as well. Most chaplains felt orphaned by their denominations. Once they entered the military, their home churches tended to forget about them and, as the pacifist movement increased during the 1930s, sometimes became openly hostile to them. Then there was the issue of promotion. A chaplain could not rise above the rank of major. Furthermore, during World War I, not a single chaplain was promoted. In addition, the chaplain school had been closed, and chaplains lacked any kind of career development structure or graduate study program. It was as if chaplains did not require additional schooling or the career evolution that other officers did. An assumption was made that a chaplain knew everything necessary to function once he joined the army. Also, no clear regulation covered chaplains' duties and responsibilities. Commanders could, and did, pile on chaplains one onerous task after another. Finally, chaplains were under constant pressure from their colleagues and commanders to consider themselves officers first and members of the clergy second. This often put them in a quandary—they were there to provide both spiritual and moral leadership for the troops, but other officers seemed to believe that their primary job was to uphold "good order and discipline." It was as if the chaplain was being asked, "Whose side are you on?" In reality, he was on both sides—he was expected to work to maintain good order and discipline, while at the same time have a close and private relationship with enlisted personnel.

The solution—which would govern the actions of chaplains until the 1950s—was the Capper Bill of June 4, 1920. This bill provided—for the first time—for a single chief of chaplains. This meant that chaplains would come under the control of one of their own. John Axton, the first chief of Army chaplains, was given the rank of colonel and invited to serve for a four-year period. His duties included a requirement to investigate carefully the qualifications of candidates for appointment as chaplains. He was also to supervise all army chaplains. This did not mean that chaplains no longer had to worry about their relationship with their commanders or with other officers. Rather, it meant that they now had some backup when that became necessary. If a commander chose to sabotage the work of his chaplain, the latter could present his complaint to a senior chaplain at a higher command, who could take care of the problem. If that person could not resolve it, there was always the possibility of having the chief of chaplains raise the issue with the appropriate senior officers in Washington.

One of the first issues that arose after Axton's appointment was whether or not chaplains should be permitted to wear their rank. The military is a very hierarchical organization, and it makes a big difference when a chaplain is immediately recognized as a captain or a major in addition to his clerical position. As early as 1918 General John Pershing had argued that the chaplains would be better off without rank devices because it would help them get closer to their men. As a result, the insignia of grade was removed from the uniform. Instead, chaplains were directed to wear crosses on shoulder loops rather than rank. Again, the issue of what a chaplain was came to the fore. Was he primarily a clergyman? If so, then rank would appear to be superfluous. If he was basically an officer, then rank was crucial, but to many in the military that would downplay his importance as a representative of his church.

In 1926 the army's chief of staff decided to resolve this issue by restoring rank. In addition, the Congressional Act of 1926—for the first time in American history—guaranteed that chaplains would be given the pay and allowances of their respective grades, up to and including the rank of colonel.[64] In addition to giving them a better bargaining position vis-à-vis other officers, the restoration of rank played an important role in helping chaplains assist their men. Despite this step forward, however, chaplains remained the only branch without a general as chief. Furthermore, since the promotion of chaplains was slower than that of physicians, many had the impression that religion played a very minor role in the military. It would be some time before this matter was rectified.

A second result of the creation of the office of chief of chaplains was that chaplains were better trained, equipped, and administered. Axten also succeeded in having the number of chaplains increased—especially among the reserves—an action that would pay major benefits when World War II broke out.

Equally important—in the minds of many chaplains—their duties were more clearly defined:

> Chaplains will be employed on no duties other than those required of them by law, or pertaining to their profession as clergymen, except when an exigency of the service . . . shall make it necessary. Chaplains are not available for detail as post exchange officers or as counsel for the defense in courts-martial.

In practice, this meant that the chaplains avoided a variety of extraneous tasks. As the same source noted, "The chaplain was freed from serving as canteen officer, postmaster, and athletic officer."[65] No longer would they have to worry about being assigned every task that no one else wanted. Their job was to provide spiritual advice, motivation, and moral support, regardless of the religious orientation of the individual concerned.

While many jobs the chaplain had traditionally been saddled with came to an end, others were added—indeed, they would become very important among his various duties. For example, like Soviet and East German political officers, the chaplain would become something of a liaison officer with social organizations. He worked directly with organizations such as the Boy Scouts, the Red Cross, the Salvation Army, the Community Chest, and the PTA program. In addition, given continuing concerns about sexual diseases and alcoholism, chaplains increasingly were expected to provide sexual hygiene classes. Indeed, "many a pragmatic commander judged the efficiency of his chaplain on the rise and fall of VD statistics in the unit."[66] Religion was fine, but issues such as venereal disease and alcoholism had a direct impact on the unit's effectiveness and thereby on the commander's chances for promotion. To be fair, these also had a direct bearing on the unit's combat readiness.

In addition to lectures on the issues noted here, chaplains continued to speak on matters directly related to the political socialization of the troops. A chaplain might offer a talk on some aspect of American history, or he might deal with local geography or the nature of the American political system. "Chaplains lectured and taught classes on citizenship and patriotic themes, current events, and sociological subjects with religious references and illustrations sprinkled amply throughout." Or as another source put it, "On the ground, much of the chaplain's time was spent in pep talks reminding GIs of the values and principles that kept America strong."[67] It is hard to communicate just how important these chaplains were to the combat efficiency of a unit. As the same source quoted previously noted,

> Just the appearance of a chaplain provided inspiration and reinforced allegiance to the code of conduct and faith in God. When problems arose over leadership, food and long tours, the chaplain would be heard affirming our role in the war, reminding the troops that "America is at war to defend freedom."[68]

Indeed, this continued to be a major part of the chaplain's job any time troops went into combat.

Efforts were also made to upgrade the quality of chaplains. In 1926 the army decided that to be considered for an appointment as a chaplain, an individual must first have the approval of his denomination. Should that approval be withdrawn, the individual would lose his commission. Candidates who were nominated by their denomination then went before a board, which would select the best man for the job. In the process, it was made clear to these men that their primary task would be to provide worship services and other religious observances for the greatest number of military personnel. In essence, this reaffirmed the old idea from the Civil War—that an effective chaplain is one who does not permit his own theological beliefs to get in the way of serving those in the armed forces. An applicant who gave the impression that he would not be able to deal effectively with those of other faiths would not be accepted. When soldiers or sailors needed spiritual or moral help, they had to be able to rely on the chaplain at hand, not search around for one of their own denomination—especially since such a chaplain might not be available.

Another innovation introduced into the chaplain corps was the idea of using census data to determine how many chaplains of each religious group should serve at any one time. Thus, based on the religious census of 1916, "98.2 percent of any authorized chaplains were determined by each group's membership figures. The remainder were divided up among churches too small to qualify for a chaplain space in the army."[69] These quotas would be revised in the future each time there was a new census.[70]

Another issue of major concern to the chaplains was the creation of a special training school. Newly commissioned chaplains had to be taught how to function in the military. In November 1919 a board of five chaplains—including three who had been active in training during World War I—met to consider the matter. The group recommended the creation of a permanent school that would conduct a five-month course twice each year. (The schools created during the war were only temporary.) It also recommended that an advanced course be developed later. The main argument was that the military had to set up schools to train civilian professionals (chaplains, doctors, dentists, and lawyers) to become military professionals. Not knowing how to salute or the nature of military customs undermined the image of chaplains in the eyes of professional officers. The 1920 Act also authorized the assignment of officers to civilian schools as students. Between 1923 and 1941, twenty-five chaplains studied in civilian institutions.[71]

On January 28, 1920, a school for chaplains opened at Camp Grant, Illinois, with a student body of fifteen. Among the various things the chaplains were taught were how to salute correctly, wear the uniform, and march, as well as details about service customs, the chain of command, and the use of special equipment. Chaplains were also instructed in the importance of reaching out to other faiths because they were responsible for men of all faiths. The problem was that it soon became evident that it would be difficult to get chaplains released to attend the school. Why? The situation was similar to that which would exist in the Wehrmacht with regard to releasing officers to attend NSFO school: "Commanders were often reluctant or unwilling to be without a chaplain in their unit for the number of weeks or months required to attend the school."[72] While some officers and commanders resented the chaplain's presence, others were beginning to learn that chaplains carried out a vital function in keeping morale up and the unit combat ready.

One of the greatest challenges the chaplaincy would face came during the 1930s, as many of the chaplains' civilian coreligionists turned to pacifism. Increasingly, the chaplains began to look like outcasts—individuals who had forsaken their churches to serve in a bloodthirsty and unnecessary military. As a civilian pastor put it in 1930, "The position of Army chaplain is in reality a wicked anachronism, and should be abolished. . . . There is no more justification for being a chaplain in the army or navy, than there is for being a chaplain in a speakeasy."[73] The underlying assumption was that civilized nations should not resort to war and that by serving in the army or the navy, chaplains were promoting state-sanctioned killings. Needless to say, this had an important impact inside the military. A number of senior military officers came to believe that since the chaplains were not wanted by their home denominations, the overall number of chaplains should be cut. The tenor of the times was such that many—both in the army and on the outside—believed that the only men who volunteered to become chaplains were either individuals who were not qualified for the life of a civilian clergyman or those who had gotten into trouble of some kind.

In evaluating this period, it is important to keep in mind that the chaplain corps was going through considerable change. It not only survived—at a time when many thought it might be on its last legs—it also grew in tasks and numbers. For example, the army was told to provide chaplains for the Civilian Conservation Corps, an experience that taught chaplains a lot about the thinking of a generation of Americans who would soon find themselves in uniform. In addition, the reserve chaplain corps, which had been built up between the wars, would also serve the military well during World War II. Although only one chaplain out of ten had been either a reserve or regular officer prior to World War II, this one-tenth would provide the basis for the many thousands who would serve in one of the largest conflicts ever fought.

World War II

At the time of the Japanese attack on Pearl Harbor, there were 140 chaplains in the regular army, 298 in the national guard, and 1,040 in the reserves, for a total of 1,478. To get some idea of how this would change, consider the following: In 1939 there were 36 Catholic chaplains in the army and 19 in the navy. By 1943, there would be a total of 3,000 Catholic chaplains in the military. By the end of the war, a total of 9,117 chaplains would serve in the army.[74] And this did not include the many thousands of civilian clergy who were sent overseas at various times to preach to the troops and lend their moral support. Clearly, this was an explosion in numbers beyond anything anyone could have imagined in the mid-1930s.

Needless to say, with the increased need for chaplains, a means had to be found to train them. The training school utilized previously by the army had been closed down in 1928. It hadn't been necessary, given the few chaplains on active duty—although training classes had been provided for reserve and guard chaplains. As a result, on February 2, 1942, less than two months after Pearl Harbor, a Chaplain's School was set up at Fort Benjamin Harrison.[75] The first class was made up of 75 students and ran for 28 days, with courses in areas such as military organization, customs and courtesies, military law, graves registration, first aid, military administration, and chaplain activities. It also exposed the students to military topics such as marching and drilling. While the school moved to different locations during the war because of the tremendous numbers involved, a total of 8,183 chaplains were given some sense of the army and what was expected of them once they went on active duty, a far cry from the days when the newly commissioned chaplain had to learn such basic things as how to salute, wear his uniform, and relate to the chain of command once he came on duty.[76]

The tasks carried out by a chaplain in World War II continued to combine a concern for spiritual matters with an effort to ensure that the soldiers, sailors, and Marines were prepared to do their duty, even if it meant (as it often did) putting their lives on the line. From a military standpoint, this was political socialization at its best. Justifying the resources that went into building chapels throughout the army, Quartermaster General Edmund B. Gregory wrote, "No matter how well a man is fed and clothed or trained, he cannot be a real soldier unless he has within him a sincere belief in the way of living of the nation which he represents. Nothing will contribute more to that belief than the opportunity for every man to worship as he chooses."[77] And to avoid any sign of religious preference, "Army Regulations provided that unit chapels should be available for services of all faiths, and that there should be nothing in the way of decoration peculiar to any one religion or offensive to others."[78] In short, if the construction of chapels would stiffen the willingness of soldiers to fight and die for their country, then the investment made sense. Former chaplains themselves have admitted that they put considerable emphasis on patriotism in dealing with soldiers. For example, Chaplain Leonard made the following observation in discussing the importance of fighting with a potential conscientious objector: "I gave him all the classic arguments for the legitimacy of a just war. I reminded him that we had been attacked. I pictured as vividly as I could the consequences to us of an Axis victory."[79]

Admiral Nimitz made a similar point when he commented on the importance of chaplains.

> My own esteem for the chaplains is not so much based on deeds of valor as it is of appreciation for their routine accomplishments. No one will ever know how many young men were diverted from acts of depression by a heart to heart talk with the "padre." . . . By his patient, sympathetic labors with men, day in day out, and through many a night, every chaplain I know contributed immeasurably to the moral courage of our fighting men. None of that appears in statistics. . . . It is for that toil in the cause of God and country that I honor chaplains most.[80]

Crosby provided a similar example from the Japanese attack on Pearl Harbor when he quoted Chaplain William Maquire, who in the midst of battle said, "Keep our spirits high and carry on with all our heart and soul in this fight for victory."[81]

Closely associated with their religious/political socialization role was their function as counselors. Indeed, from the perspective of a commander this was one of the most important things they did. Chaplains were assured of the privacy of their conversation, and on more than one occasion, the ability of a soldier or a sailor to talk to someone privately helped him deal with his personal problems in a way that kept him militarily effective.

> In the commander's view of the chaplain's role, nothing was more important than that of counselor. Whether conducted in an informal setting as the chaplain visited about the post, in the more formalized setting of an office, or in the hospital or stockade, the chaplain was sought for counsel by enlisted men and officers for a great variety of problems. The Chief of Chaplains repeatedly emphasized the responsibilities of chaplains in protecting the sacred relationship of the confession or the privileged communication made to a chaplain in counseling.[82]

In 1942 alone, chaplains had an average of fifty-three personal conferences every day. In the United States 12 percent of all army personnel consulted a chaplain in the course of one year, while overseas, the number rose to 25 percent. Chaplains had to deal with everything from homesickness, suicidal feelings, marriage problems, and alcohol to "Dear John" letters. A soldier or sailor who was worried about problems at home would be less likely to perform as well on the battlefield.

Chaplains also often found themselves dealing with a host of other problems. For example, they continued to give "VD talks," warning soldiers and sailors of the moral and social consequences of promiscuous sexual behavior. Often, they were put in charge of entertainment, recreation, and sports facilities. Indeed, many chaplains found sports to be a useful vehicle for getting closer to the men and women they served. Chaplains were also the military's liaison officer with groups such as local churches, the Red Cross, the YMCA, Special Services, the USO, and the Salvation Army, as well as other civilian groups that tried to help raise the morale of those serving in the armed forces.

Chaplains played an important role in helping to prepare men for battle. Frequently, soldiers and sailors knew they would be going into combat the next day, so the chaplain went around organizing prayer services for men—many of whom were not especially religious but who recognized the value of religion when their lives were on the line. Chaplain O'Callahan described the prayers he organized aboard the USS *Franklin* the evening before the ship went into combat.

> In our prayers before entering combat, these boys and I had not asked to escape unscathed, to come out alive; we had asked Divine Assistance to do a good job for God and country. We had reminded ourselves that, should death come, in whatever form, it would be a happy death if we died in the friendship of God.[83]

Or as famed wartime cartoonist Bill Mauldin put it, "I have a lot of respect for those chaplains who keep up the spirits of the combat guys. They often give the troops a pretty firm anchor to hang onto."[84] The bottom line was that chaplains were crucial in preparing men for battle, in much the same way that political officers and commissars would be.

Probably the strangest request put to any chaplain during the war was that which befell Chaplain James O'Neill. On December 14, 1944, General George Patton called O'Neill into his office at Third Headquarters in Nancy, France. At the time, Patton was frustrated by the wet weather that was holding up the Third Army's attempt to relieve surrounded troops at Bastogne. Patton told Chaplain O'Neill that he wanted him to write a "prayer for good weather. I'm tired of those soldiers having to fight mud and floods as well as Germans. See if we can't get God to work on our side." Needless to say, Chaplain O'Neill was a bit taken aback and told Patton that "it isn't a customary thing among men of my profession to pray for clear weather to kill my fellow men." Patton refused to back down and sent the chaplain off to pen an appropriate prayer. After consulting with his Anglican colleague, O'Neill (who was Roman Catholic) came up with the following:

> Almighty and most merciful Father, we humbly beseech Thee, of Thy great goodness to restrain these immoderate rains with which we have had to contend. Grant us fair weather for Battle. Graciously hearken to us as soldiers who call upon Thee that, armed with Thy power, we may advance from victory to victory, and crush the oppression and wickedness of our enemies, and establish Thy justice among men and nations. Amen.[85]

The day after the prayer was issued to the troops, the weather cleared up and re-mained good for the next six days. Obviously, Patton believed in both God and the importance of the chaplain in convincing the Almighty to provide the Third Army with the weather it needed to get the job done. As he noted later, he had a super-vising chaplain "with powerful influence in heaven."[86]

In dealing with problems, emphasis was placed again on the need for a chaplain to rise above his own denominational or theological beliefs. Indeed, if any thread runs through the history of the chaplains' corps from the Civil War to the present, it is the idea that chaplains serve their troops, regardless of what their religious orientation might be. As one former chaplain noted, "A chaplain, in the total institutional envi-ronment of the military, serves the entire military society rather than those of his own denomination. This is perhaps the most important difference between any in-stitutional chaplaincy and the parochial ministry to a congregation of a particular denomination."[87] Gushwa provided a practical example when he noted the following incident, "'This is our chaplain,' said a young paratrooper at Fort Bragg, North Car-olina. The chaplain was Catholic, the parents of the soldier Methodist, and he him-self was not affiliated with any church. But the chaplain was still his chaplain."[88]

It is also worth noting that chaplains were again in the front lines and in many cases performed heroically. They were in prison camps along with their compa-triots and ministered to them. For example, when Corregidor and Bataan in the Philippines fell, twenty-one chaplains went into Japanese prison camps. Among the many examples of heroism during World War II, few compare with the actions of Chaplain Joseph T. O'Callahan, a Jesuit priest who was serving on the USS *Franklin* when it was hit by a Japanese kamikaze attack. Instead of limiting him-self to providing spiritual counseling to dying and wounded sailors, O'Callahan organized and led fire-fighting crews, helped throw live (and hot) ammunition over the side, and constantly moved from one of the most dangerous sections of the ship to another in an effort to save as many sailors as he could.[89] In recogni-tion of his heroism, he was awarded the Medal of Honor. Many other examples could be provided of cases where chaplains risked their lives to bring in the wounded while under fire from snipers, machine guns, and mortar shells.

As an example of the key role the chaplains played in the front lines, consider the following: "The chaplain branch was third in combat deaths on a percentage basis behind the Air Forces and the Infantry. From Pearl Harbor to September 30, 1945, there were a total of 478 casualties among army chaplains."[90] This high casualty rate was a result of the attitude on the part of chaplains that it was their task to "be there." If soldiers, sailors, or Marines were wounded or dying, it was the chaplain's job to be with them to provide them with whatever assistance they needed.

Catholic chaplains . . . felt that their place was with the dying, and many of them were killed while giving the last rites. Protestant chaplains often felt that faith in the Lord gave men courage to face anger, and being "up front" was for them a logical extension for practicing what they preached. Jewish chaplains were less numerous and assigned to higher headquarters and tended to be less exposed to combat conditions.[91]

In evaluating the overall performance of chaplains during World War II, it is important to note that in comparison to the First World War, they were much better organized and they were given far more latitude to perform their duties. Commanders still exerted considerable influence, but their ability to treat chaplains as second-class citizens was gone. The chaplain had become an officer and a clergyman, a person to be reckoned with. As far as his value to the military effort was concerned, military commanders felt that chaplains played an important role—not so much because they were religious leaders, although some commanders may have found that significant, but because chaplains contributed significantly to accomplishing the task of winning the war. Indeed, the Japanese were so concerned about chaplains' influence that, at first, religious services were not permitted in Japanese prison camps and later were allowed only under the proviso that all sermon texts had to be submitted in advance to the camp commandant. The Japanese were apparently afraid that religious services would have the effect of strengthening the prisoners' willingness to resist.

The Postwar Period

One of the first important actions that had a direct impact on the chaplain corps was the creation of a joint agency to consider matters of interest to all chaplains—regardless of service. On July 18, 1949, Washington established the Armed Forces Chaplain Board within the Office of the Secretary of Defense. It was made up of the three chiefs of chaplains and one additional chaplain from each of the three services (the Marine Corps utilizes Navy chaplains, as does the Coast Guard). The idea was that this group would make it easier to make decisions that might impact on chaplains in all of the services. For example, there is the problem of procurement. In many areas chaplains in one service utilize the same equipment as those in other services, so why not work together in order to keep prices down? The same was true with training standards and programs. Although specific service differences would always exist, in a lot of cases chaplains could train together.

At the same time, chaplains were given increased responsibility for morality, character building, and patriotism. Toward this end, commanders were ordered to schedule periods in the training schedule so that chaplains could address these issues. As a consequence, chaplains intensified their work in these areas. This led the navy to develop a program that would "be a continuous process for all personnel in whatever area of service they may happen to be, whether in Recruit Training, Service School, Officer Candidate School, service with the Fleet, or on duty on a foreign station."[92] In fact, this program went so far that from 1952 to 1956 a number of chaplains were assigned to develop character education curricula for the various types of schools, stations, ships, and commands.

Despite concerns on the part of some people over the chaplain's involvement in character building, the issue did not go away. In Korea, for example, American chaplains became seriously involved in this process when they discovered that Americans appeared to lack the moral fiber to stand up to the communists once they had been captured. This behavior contrasted, for example, with the British Gloucestershire Regiment, which had also been overrun and captured by the North Koreans. Not one of them went over to the communists, even though their lives in prison had been very hard. Many believed that this was the result of the British regiment's strong religious faith—at least, enough people believed it so that the American military came to the conclusion that moral education was a priority.

The problem with the moral education program was that many in the military—including a large number of chaplains—worried that this combination of moral and civic education with religion was inappropriate. They were concerned that it could lead to challenges to the chaplain corps from critics who argued that it was a violation of the establishment clause of the First Amendment. As a consequence, chaplains gradually became less and less involved in it.

Meanwhile, the position of chaplains vis-à-vis other officers had improved considerably over what it had been in years past. Take, for example, the case of Claude Newby, a Mormon chaplain, who served two tours in Vietnam. Chaplain Newby's duties led him to visit a variety of military posts in Vietnam—always by helicopter. At one point, he was dependent on a Captain Root for his helicopter flights. Root was an atheist and resented Newby's presence in the country. Root would give Newby false information about flight schedules. This continued until finally Newby confronted Root. Root exclaimed, "Chaplain Newby, I've been an atheist all my life. And as far as I am concerned, you are a bad influence on the troops, and I'm duty bound to protect them from you. I'll do everything I can to make you miss flights and otherwise hamper your activities." Newby responded, "I will not be bumped, and if you give me false information about flights or interfere in any way with religious support, you and I will be standing before the battalion commander."[93] That ended the issue. Root understood that he would lose such a confrontation. Chaplains had to be taken seriously, a far cry from the situation Newby would have found himself in fifty years previously.

Just as was the case with Chaplain O'Callahan in World War II, chaplains in Vietnam also used their offices as a means to strengthen the fighting spirit of the men they served with. In response to a question—what the men prayed for—one Baptist chaplain responded, "Courage, protection, life, victory! Many ask God to take sides in the war in Vietnam. They say that God is on our side. . . ."[94] From a military standpoint, such prayers can work wonders when it comes to fighting. They give soldiers and Marines a reason to fight and kill. And unlike the case with Chaplain Newby cited previously, many soldiers and officers recognized their value. Consider the following:

> During the Vietnam conflict, a chaplain accompanied his unit on patrol through the dark, thick, rain-soaked foliage of a remote jungle somewhere in Southeast Asia. As

the patrol cautiously approached the crest of a hill, the men paused, sensing danger. They all recognized that they could be walking into an ambush.

A small squad, it was decided, would advance. The rest would stay back. As the chaplain began to move out with the advance party, the young officer in charge stopped him. "Chaplain, you stay. We can't afford to lose you."[95]

If Vietnam did nothing else, it clearly put chaplains back into the character guidance business in a big way. They conducted thousands of such classes all over the world—and this included classes for officers as well as enlisted men and women. Problems with alcohol, drugs, and race relations had reached epidemic proportions. Discipline was out of control in many parts of the world. The military also seemed to have lost its sense of purpose. It was clear that chaplains would have to spend increasing amounts of their time working on morale, motivation, and political socialization.

And chaplains responded. The chiefs of chaplains mounted a major effort to turn things around. To improve race relations, a drive to recruit black chaplains was undertaken and affirmative action plans were instituted. A chaplain set up the first human relations council in 1971. Racial Harmony Workshops were organized all over the world. Chaplains also became the first de facto equal opportunity representatives in their units. Their involvement in dealing with drugs was especially extensive. By June 1971, the army had assigned eighty-two chaplains and eighty-one chaplain assistants to be drug counselors.[96] Chaplains conducted day-long workshops throughout the military on drug and alcohol abuse. As the Vietnam War came to a close, the drug problem seemed to lessen. By the 1980s most chaplains were out of the drug counseling business.

Drug counseling, however, was quickly replaced by even more emphasis on race relations, now called multiculturalism. Chaplains sponsored gospel music workshops, and they were the primary promoters for events such as Martin Luther King Day, Black History Month, and other similar cultural events. In fact, in the army, the chief of chaplains made race relations a priority issue. "The Army chaplaincy must bring to bear the resources of religious faith and work within this framework to alleviate the situation." Chaplains also became involved in the Personal Effectiveness Training program, an effort to teach officers how to deal better with enlisted personnel. Furthermore, chaplains began to spend even more time working with families as the divorce explosion hit the military.[97] Finally, as this book goes to press, chaplains have been given another assignment—to deal with civil affairs if and when the military occupies an area or is stationed outside the United States. In this sense, the chaplaincy has expanded far beyond the parameters of the spiritual leader he was during the Colonial period.

Like the rest of the country, the chaplain corps was also becoming increasingly diverse. During the American Revolution, only seven denominations were represented among the chaplains serving in the Army. During World War II, chaplains served from 40 different faith groups. By 1987, 109 denominations were represented in the Army. The situation was the same in the Navy. In August 1945 there were 25 denominations in the Navy. By October 1983 these groups numbered 83 and by 1987 they were up to 90.[98] Indeed, the situation had reached the point that by 1999, witches were even demanding recognition as a faith group and presumably expecting the appointment of chaplains from their ranks.[99]

One of the more unique issues that the chaplain corps had to face in the postwar period came in Operation Desert Storm. Saudi Arabia was a very conservative Muslim country—one that was not open to the presence of Christian chaplains, not to mention those of the Jewish faith. It was also a country in which alcohol was forbidden. However, wine played a central role in the liturgy of a number of religious faiths. In an effort to avoid conflict with the host authorities, instructions were issued that read "Chaplains are authorized to possess such items as are necessary to conduct religious services and to use such items with discretion in the provision of religious and spiritual programs." Concerning their insignia, which identified them as either Christian or Jewish chaplains, they were told that they could wear either the cross or the tablets as long as they were in U.S. controlled areas. Once they left U.S. areas, however, they were expected to remove their chaplain's insignia.[100]

In addition to holding religious services, chaplains also took care of burial details, handled refugees and POWs, and provided other types of humanitarian assistance as necessary. To quote one source, "The primary focus is on helping military personnel stay focused on their duties."[101] It is also worth noting that one thing that has not changed is the importance of chaplains gaining the respect of the soldiers, sailors, airmen, or Marines they serve. As another source noted during the Gulf War, "When they have the trust and respect of the troops, they are better able to nurse them through boredom, loneliness and depression and prepare them to deal with death—their own or that of an enemy at their hands."[102]

Conclusion

One could argue that, by the time this book went to press, most of a chaplain's time was devoted to ancillary duties, those related to morality and values, and only partly related to the duties carried out by chaplains of the Revolutionary War period. Religion was still important—although with the explosion in faith groups, even its nature had changed. Muslim, Jewish, and a myriad of new Protestant denominations were present in all of the services.

If any duties remained constant throughout the years—in addition to religious services—they were personal counseling and the expectation that the chaplain would be present on or near the battlefield in time of war to inspire and console those who served in the armed forces.

Chaplains have always been available to deal with the serviceman's or
-woman's personal problems. Concerning chaplains' presence in combat, little
has changed since the beginning of this country's history. Chaplains were expected to be there, and regardless of whether it was World War I, World War II,
the Korean War, Vietnam, or Desert Storm, chaplains have been there and at
times have performed acts of heroism. Since the Civil War, the military has also
expected chaplains to serve the needs of all of the soldiers or sailors who serve
with them. This is even more true today, as the armed forces become more diverse. In World War II it was primarily a matter of Protestants, Catholics, and
Jews learning how to deal with officers and enlisted personnel from their various groups. Now, however, with the tremendous expansion in the number of
different faith groups represented in the chaplain's corps, the issue is much
more complex. A priest or minister must know how to relate not only to a Jew
but to a Muslim or any number of groups.

The other factor that arises again and again in the evolution of the American
military chaplain is the military establishment's expectation that chaplains will
work to inculcate those values most supportive of the armed forces in the minds
and hearts of the troops with whom they serve. Chaplains would not be expected
to hold daily indoctrination lectures, as was the case with political officers, but
they were tasked with providing a religious belief that would translate into convincing soldiers, sailors, and Marines to die for their country. This point was made
by a couple of critics of the chaplains corps when they observed that "military religion, in addition to its direct legitimization of the military enterprise, also legit-
imates the latter by advocating values that are functional to soldierly performance
and by de-emphasizing values that are potentially dysfunctional."[103] These two
scholars' analysis of the materials utilized by chaplains showed that the linkage between God and country was stressed by chaplains in carrying out their work. My
own experience with military chaplains over the years, as well as material in chaplains' memoirs, suggests that this has always been a prime concern. In this sense,
the chaplain carries out a very important military function. Soldiers or sailors are
not of much use if their morale is down or if they lack discipline. By raising
morale, fighting spirit, and devotion to country, and by providing a justification
of the trials and tribulations of military service, chaplains carry out a very important mission. For this reason, most people in the military consider chaplains' services to be indispensable.[104]

Conflicts of conscience would also remain a problem for chaplains. As recently as 1997, for example, a group of chaplains took on the Defense Department because of its effort to avoid becoming involved in what it saw as a political debate. U.S. Roman Catholic bishops urged their pastors to fight against
partial-birth abortion. "Specifically, the bishops hoped that a substantial contribution to the goal of overturning President Clinton's veto of the Partial-Birth
Abortion Ban could be accomplished by asking parishioners to take part in the
'Project Life Postcard Campaign.'" Meanwhile, the Pentagon ordered chaplains

not to get involved in this project. A Roman Catholic chaplain ignored this order and drew the ire of the Defense Department. This priest then took the issue to court. He won. The court called it censorship.[105]

In terms of authority relationships, the chaplain appears to have finally arrived at the point where he or she (women can now become chaplains) is taken seriously by the rest of the command structure. There will always be people who resent the presence of chaplains because soldiers or sailors can take their complaints against the hierarchy to the chaplain. But all military personnel—including the commander—recognize that they cannot ignore the chaplain or foist a lot of unwanted jobs on him or her. This in itself marks a major change in the relationship that was present during much of the time period covered by this chapter. To quote one chaplain,

> When I enlisted in World War II, the chaplain was tolerated. The commanding officer either turned him loose to do what he could or wanted to do—or tried to use him as a flunky and recreational officer. . . . But today the commander recognizes the chaplain as an important member of the staff and expects him to develop a spiritual program for the troops.[106]

The chaplain's concern with morale, motivation, and political socialization will be shared by all of the other military entities in this book. Some will avoid the religious orientation and instead substitute an ideology such as Marxism-Leninism as a vehicle for motivating soldiers and sailors. Others, such as Cromwell's army, will rely on religion in much the same way that the American chaplain did to motivate the troops for battle. In this sense I have in mind not only chaplains being present on the battlefield—as they generally were—but also convincing soldiers to train for combat. This was indirect, in the case of a chaplain, but very direct when it came to a political officer or commissar. If the unit performed badly, political officers and commissars would suffer personally.

All of the different individuals discussed in this book were concerned with counseling personnel. It is common in the U.S. military to hear the expression "Tell it to the chaplain." However, counseling personnel was also a matter of concern to both commissars and political officers. Individual counseling took up a considerable amount of their time. Soldiers needed someone to talk to about the stress of military life, as well as their personal problems. In the absence of a chaplain, political officers or political commissars were called upon to carry out this function.

Finally, in one area all nonchaplains differed from the American experience: their relationship to the commander. In some instances, the officer carrying out the motivation, morale, and political socialization function actually had control over the actions of line officers, including the commander. In other cases, he was subordinate but still in the chain of command. This contrasts sharply with the case of American chaplains, who labored over one hundred years just trying to get to a position where they were not subject to the arbitrary actions of their direct military superiors.

Notes

1. Much of the material in the next several sections is drawn from Roy J. Honeywell, *Chaplains of the United States Army* (Washington, D.C.: Office of the Chief of Chaplains, 1958), 11.
2. Honeywell, *Chaplains of the United States Army,* 17.
3. Honeywell, *Chaplains of the United States Army,* 28.
4. As quoted in George Houston Williams, "The Chaplaincy in the Armed Forces of the United States of America in Historical and Ecclesiastical Perspective," in Harvey Cox, *Military Chaplains* (New York: American Report Press, 1973), 18.
5. Cox, *Military Chaplains,* 18.
6. Joel Headley, *Chaplains and Clergy of the Revolution* (New York: C. Scribner, 1864), 59, as cited in William Dickens, *Answering the Call: The Story of the U.S. Military Chaplaincy from the Revolution through the Civil War,* Dissertation for the Southern Baptist Theological Seminary, 1998, 79.
7. Honeywell, *Chaplains of the United States Army,* 30. The chaplain of this time period was quite different from what we have today. They were not as disciplined and certainly not pluralistic in their view of the religious role. Indeed, one writer noted that in the UK—which served as the model for the revolutionary army—"In 1764 a chaplain was killed fighting a duel in Epping forest." John L. Rand, *The Development of the Military Chaplain in Great Britain* (June 1968), 13. Beyond a general belief that chaplains shared responsibility for the unit's morale and that it was up to them to help motivate the troops, their roles were not well defined. Much depended on the unit commander and what he believed a chaplain should be doing.
8. Honeywell, *Chaplains of the United States Army,* 58.
9. Honeywell, *Chaplains of the United States Army,* 38.
10. Dickens, *Answering the Call,* 20.
11. Gordon Taylor, *The Sea Chaplains* (Oxford: Oxford University Press, 1978), 177.
12. Honeywell, *Chaplains of the United States Army,* 74.
13. Honeywell, *Chaplains of the United States Army,* 31.
14. Honeywell, *Chaplains of the United States Army,* 35.
15. Williams, "The Chaplaincy in the Armed Forces of the United States . . . ," 24.
16. The tendency to cut back on the number of chaplains in the aftermath of a war was not unique to the U.S. military. In 1811, for example, Wellington had only "one chaplain in his whole army." John Smyth, *In This Sign Conquer* (London: Mowbray, 1968), 31.
17. Honeywell, *Chaplains of the United States Army,* 76
18. As cited in Honeywell, *Chaplains of the United States Army.*
19. Williams, "The Chaplaincy in the Armed Forces of the United States . . . ," 25.
20. Williams, "The Chaplaincy in the Armed Forces of the United State . . . ," 26.
21. From 1818 to 1838 Congress did not authorize chaplains in the army. As noted previously, during that time, "some posts hired their own, while commanders at others marched their men to services in nearby villages." Edward M. Coffman, *The Old Army: A Portrait of the American Army in Peacetime, 1784–1898* (New York: Oxford University Press, 1986), 178.
22. Coffman, *The Old Army,* 78. The idea of using chaplains as teachers was not specifically American. They also carried the load in this area in the British military. Smyth, *In This Sign Conquer,* 62–63.

23. Coffman, *The Old Army*, 180.

24. Rollin W. Quinby, "Congress and the Civil War Chaplaincy," *Civil War History* 10, no. 3 (1964): 246.

25. Coffman, *The Old Army*, 176.

26. Percival Lowe, *Five Years a Dragoon* (Norman: University of Oklahoma Press, 1965), 24, 146, 150.

27. Williams, "The Chaplaincy in the Armed Forces of the United States . . . ," 27–28

28. Albert Isaac Slomovitz, *The Fighting Rabbis* (New York: New York University Press, 1999), 11. See also Charles F. Pitts, *Chaplains in Gray: The Confederate Chaplain's Story* (Nashville: Broadman, 1957).

29. Warren B. Armstrong, *For Courageous Fighting and Confident Dying: Union Chaplains in the Civil War* (Lawrence: University Press of Kansas, 1998), 2.

30. Slomovitz, *The Fighting Rabbis*, 12.

31. Alan K. Lamm, *Five Black Preachers in Army Blue, 1884–1901* (Lewiston, N.Y.: Edwin Mellen, 1998), 64.

32. Dickens, *Answering the Call*, 2.

33. Lamm, *Five Black Preachers in Army Blue*, 66.

34. Rev. William W. Lyle, *Lights and Shadows of Army Life* (Cincinnati, 1865), 117–118, as cited in Armstrong, *For Courageous Fighting and Confident Dying*, 21.

35. As quoted in Honeywell, *Chaplains of the United States Army*, 103.

36. Dickens, *Answering the Call*, 105.

37. Honeywell, *Chaplains of the United States Army*, 137.

38. As quoted in Dickens, *Answering the Call*, 117.

39. One of the best discussions of this situation is in Dickens, *Answering the Call*, 68–75.

40. Honeywell, *Chaplains of the United States Army*, 148.

41. Bell Irwin Wiley, *The Life of Billy Yank: The Common Soldier of the Union* (Indianapolis: Bobbs-Merrill, 1951), 267.

42. Much of what follows for the period up to 1920 is based on Earl F. Stover, *Up from Handymen: The United States Army Chaplaincy, 1865–1920* (Washington, D.C.: Office of the Chief of Chaplains, 1977), 2.

43. William Corby, *Memoirs of Chaplain Life* (New York: Fordham University Press, 1992), 57.

44. As quoted in Dickens, *Answering the Call*, 128.

45. Armstrong, *For Courageous Fighting and Dying*, 17.

46. Corby, *Memoirs of Chaplain Life*, 185–186.

47. Williams, "The Chaplaincy in the Armed Forces of the United States . . . ," 41–43.

48. Williams, "The Chaplaincy in the Armed Forces of the United States . . . ," 43.

49. A lieutenant junior grade, or LT(jg), is the equivalent of an army first lieutenant.

50. Stover, *Up from Handymen*, 2. See also Ewin S. Redkey, "Black Chaplains in the Union Army," *Civil War History* 33, no. 4 (December 1987): 331–350.

51. Coffmann, *The Old Army*, 323.

52. Stover, *Up from Handymen*, 3.

53. Integrating Roman Catholic chaplains into a Protestant-dominated military was not easy. Many resented their presence. This was as true of the United Kingdom as it was of the United States. Roman Catholic chaplains were not introduced by London until 1836 and then they were not placed under the supervision of the chaplain-general, as was the case with Anglican and Presbyterian chaplains, but under the direct control of the War office.

They were not given regular commissions as officers until 1859. Smyth, *In This Sign Conquer*, 28, 113.

54. Coffman, *The Old Army*, 391.
55. Stover, *Up from Handymen*, 147
56. Stover, *Up from Handymen*, 148.
57. Stover, *Up from Handymen*, 186.
58. Honeywell, *Chaplains of the United States Army*, 175.
59. This battle raged in other militaries as well. For example, during the Korean War the Canadian military began giving chaplains rank—rather than a honorary rank. Beginning in October 1950 all serving chaplains and newly accepted chaplains in the Army and Air Force were commissioned as officers. Albert Fowler, *Peacetime Padres* (St. Catherine's, Ontario: Vanwell, 1996), 77.
60. British and New Zealand chaplains were expected to play the same role at this time. Smyth, *In This Sign Conquer*, 170; Jack S. Harker, *Soldier, Sailor, Priest* (Auckland, N.Z.: Challenge Communications Foundation, 1992), 57.
61. Honeywell, *Chaplains of the United States Army*, 193.
62. Smyth, *In This Sign Conquer*, 165. The bottom line in all armies using chaplains was, as Smyth put it, "There is little doubt that it was the chaplains who shared the discomforts and dangers of the front line with the units to which they were attached who had the most influence over the men." Ibid.
63. Discussion of the role and functions of the chaplain from 1920 to 1945 draws heavily on Robert L. Gushwa, *The Best and Worst of Times* (Washington, D.C.: Office of the Chief of Chaplains, 1977).
64. Despite the tendency of the American chaplaincy to follow the lead of the English, it is worth noting that, as early as 1858, the English had given their chief of chaplains (or chaplain-general to the forces) the equivalent rank of major-general. Smyth, *In This Sign Conquer*, 113.
65. Gushwa, *The Best and Worst of Times*, 9.
66. Gushwa, *The Best and Worst of Times*, 13.
67. Gushwa, *The Best and Worst of Times*, 25, and Robert L. Gushwa, "Keeping the Faith," *American Legion Magazine* 139, no. 3 (September 1995): 30.
68. Gushwa, "Keeping the Faith," 31.
69. Gushwa, *The Best and Worst of Times*, 16.
70. The British also divided up chaplains according to population. As one source put it, "The establishment for each domination was based on one chaplain for 1100 men of that domination." Smyth, *In This Sign Conquer*, 207.
71. Gushwa, *The Best and Worst of Times*, 19.
72. Gushwa, *The Best and Worst of Times*, 19.
73. Gushwa, *The Best and Worst of Times*, 53.
74. Gushwa, *The Best and Worst of Times*, 97.
75. The British also faced the need to create chaplains' schools. The first one was set up in March 1944, although English schools would never reach the size of the American chaplain schools, as some 200 attended this first one. Smyth, *In This Sign Conquer*, 244–247.
76. For a discussion of chaplains' school, see William J. Leonard, S. J., *Where Thousands Fell* (New York: Sheed and Ward, 1995), 8–10.
77. As cited in Gushwa, *The Best and Worst of Times*, 115.
78. Leonard, *Where Thousands Fell*, 92.

79. Leonard, *Where Thousands Fell*, 16.

80. As cited in Gushwa, *The Best and Worst of Times*, 163.

81. Donald F. Crosby, S. J., *Battlefield Chaplains* (Lawrence: University Press of Kansas, 1994), 15.

82. Gushwa, *The Best and Worst of Times*, 132.

83. Joseph T. O'Callahan, S. J., *I was Chaplain on the* Franklin (New York: Macmillan, 1961), 54.

84. Crosby, *Battlefield Chaplains*, 100.

85. As quoted in George S. Patton, Jr., *War as I Knew It* (New York: Bantam, 1979), 176.

86. Crosby, *Battlefield Chaplains*, 155.

87. As cited in Gushwa, *The Best and Worst of Times*, 113.

88. Gushwa, *The Best and Worst of Times*, 112. It is worth noting that the United States was far ahead of the British in this area at this time. Pocock, an Anglican chaplain, tells, for example, of being asked to conduct services for Roman Catholics in 1939 and in the process notes how unusual such a procedure was. As he put it, "the RCs were very much apart from the other denominations in those days." Lovell Pocock, *With Those in Peril* (Worcester: Self Publishing Association, 1989), 33.

89. His story is in Father Joseph T. O'Callahan, S. J., *I Was Chaplain on the* Franklin.

90. Gushwa, *The Best and Worst of Times*, 141.

91. Gushwa, *The Best and Worst of Times*, 142.

92. Richard G. Hutcheson, Jr., *The Churches and the Chaplaincy* (Atlanta: John Knox, 1973), 149.

93. Claude D. Newby, *It Took Heroes* (Springville, Utah: Bonneville, 1998), 124.

94. Wayne Dehoney, *Disciples in Uniform* (Nashville: Broadman, 1967), 48.

95. Donald Hadley and Gerald Richards, *Ministry with the Military* (Grand Rapids, Mich.: Baker, 1992), 27.

96. John W. Brinsfield, Jr., *Encouraging Faith, Supporting Soldiers* (Washington, D.C.: Office of the Chief of Chaplains, 1997).

97. Canadian chaplains also had to deal with a significant increase in responsibility for things like character building, drugs, and alcohol during the postwar period. Fowler, *Peacetime Padres*, 96.

98. Brinsfield, *Encouraging Faith, Serving Soldiers*, 242.

99. "An Army Controversy: Should the Witches Be Welcome?" *The Washington Post*, 8 June 1999.

100. Brinsfield, *Encouraging Faith, Supporting Soldiers*, 59.

101. "The Iraq Crisis," *The Atlanta Constitution*, 19 December 1998.

102. "US Troops Seek Solace in Faith: Chaplains Say Their Services Are in Demand," *The Atlanta Constitution*, 9 February 1991.

103. Peter Berger and Daniel Pinard, "Military Religion: An Analysis of Education Materials Disseminated by Chaplains," in Cox, *Military Chaplains*, 99.

104. It is interesting to note in this regard that in chapter 12 of the *King's Regulations and Admiralty Instructions*, which deals with "Discipline, Sunday Work and the Chaplain," the chaplain is the first person to be mentioned under the term *discipline*. From the standpoint of the British Navy, the chaplain was expected to play a very important role in this area. Lovell Pocock, *With Those in Peril* (Worcester: Billing and Sons, 1989), 19.

105. Stanley T. Grip, "The Military Chaplain: An Incongruous Man?" *New Oxford Review* 7, no. 1 (1997): 22.

106. Dehoney, *Disciples in Uniform*, 67.

Part II
The Case Studies

2

Chaplains in Cromwell's New Model Army

Truly, I think that he that prays best will fight best.

—Oliver Cromwell

OLIVER CROMWELL'S NEW MODEL ARMY did not have political commissars or political officers; instead, it had chaplains. Chaplains were not new in the English military. During the early period, many clergymen served as soldiers.[1] The situation changed under King Edward I in the late thirteenth century. By the time of the Battle of Crécy in 1346, chaplains were utilized widely for the first time. From that point on, they were a regular part of the English Army. Indeed, by the beginning of the seventeenth century, the regimental colonel "was expected to have a well-governed and religious preacher in his regiment so that by his life and doctrine soldiers may be drawn to goodness."[2]

The English Civil War was unique in English history. As Smyth said, it "was definitely not a war of class hatred or greed, but a conflict of political and religious beliefs that divided every rank in a land which was socially sound and economically prosperous."[3] This meant that in addition to providing soldiers with spiritual advice, the Parliamentary Army—which were the rebels, in this case—had to convince those who served in its ranks of the justness of their cause, to persuade them that they owed allegiance to this body instead of to the king, who had been viewed as a semireligious figure in the past.

This meant that the chaplains in Cromwell's army would play a pivotal role in raising morale, motivating soldiers, and politically socializing them in the political battle between the forces of Parliament and those of the king.[4] The struggle between these two political giants was generally depicted in religious terms: a struggle between the forces of good and evil. And it was the chaplain's job to ensure that the soldiers on Parliament's side were not only combat ready but fought as hard as humanly possible. In this sense, he was far more than a spiritual adviser. His task was to convince soldiers that the war being waged was a holy one.

In contrast to traditional professional armies, when it came to motivating soldiers in Cromwell's army, pay took second place behind the soldier's belief that he was doing the "Lord's will." This was not the secularized ideology of the French or Russian revolutions, but it was one of the first cases in history when ideology played such an important role with the mass of soldiers, and those whose task it was to inspire troops occupied a central position.

Although one writer downplays the role of religion as a form of ideology—noting that the number of chaplains was limited in Cromwell's army—in fact, more chaplains were present than that writer assumed. Furthermore, without chaplains, Cromwell's success on the battlefield would have been more in doubt.[5] Indeed, the fighting spirit of Cromwell's soldiers often made the critical difference in battle. The professionals—who were primarily on the king's side—were "disappointing. . . . they were apt not only to leave the battlefield in search of plunder at crucial moments, but also to surrender positions they thought untenable." Hill provides another example when he notes that General Monck, a professional soldier, went over to Parliament's side when the latter was winning and then in 1660 changed back to the king's side when it was clear that Parliament was losing.[6] This kind of "wishy-washy" commitment to an ideal, an ideology, or a religious tenet was not the way that Cromwell's soldiers were trained. They were committed to a cause, and the army's chaplains were an important vehicle for motivating them.

Before we discuss the other roles played by the chaplains in Cromwell's New Model Army, it is important to understand the setting in which the English Civil War took place.

The English Civil War

The Civil War marked a major turning point in English history. It signified the end of the Middle Ages and the beginning of a new era. As a result, it was a transitional period. The populace was religious—the secularizing influence of the French Revolution was still to come. And regarding the military, the English Civil War marked one of the first times that efforts were made to gain the average soldier's personal commitment to a cause. In the past, most soldiers generally did not care why or whom they fought—as long as they were paid. During the English Civil War, however, men who were not professional soldiers were being asked to lay their lives on the line in support of Parliament in its battle against the king and the established Anglican Church. In most cases, these men's opponents were highly trained and well armed.

As a consequence, what soldiers thought and believed was important—so important, in fact, that a special group of individuals, whose task it was to motivate them, played a very important role. The only thing these amateur soldiers had to make up for their lack of professional expertise was their willingness to sacrifice and, if need be, lay their lives on the line for something they believed in. Professional soldiers understood very well how to fight wars, but they were trying to maximize their own personal return—in terms of pay or loot. Dying for an idea was not something that motivated them.

During the 1630s Charles I, the king of England, made a serious attempt to regain the authority that the crown had lost to Parliament in the preceding century. For him, the idea of sharing power with anyone was unacceptable. His goal was nothing less than the creation (or re-creation) of an absolute monarchy such as those that existed elsewhere in Europe. Toward this end, he and his government behaved in ways that were bound to upset Parliament—which was jealous of its power and did not want to do anything that might diminish it.

The king's men used arbitrary arrests—often followed by imprisonment—as a way of raising taxes while circumventing Parliament. (Those who did not pay what the king's men requested—even if the tax had not been levied by Parliament—would be arrested on a trumped-up charge.) Charles I also relied heavily on the courts he controlled as a means to enforce government policy.[7] Gradually, the crown's efforts to limit, if not eliminate, local autonomy became increasingly evident.

Meanwhile, from the other side, pressure for change was everywhere. Agricultural production was a prime example. With its growing population, pressure for increased production was inevitable. Yet the crown opposed such increases, by trying to keep large amounts of land under its own control (and not cultivated) and by arresting those who tried to make use of it for growing crops. Indeed, the king seemed more interested in protecting his lands than in preventing starvation, a phenomenon that hit England hard from 1593 to 1597.

Charles's attempt to reassert his personal authority over the political and economic system coincided with the Anglican Church's effort to reassert its control in the religious sphere. Until England's defeat of the Spanish Armada in 1588, the crown depended heavily on support from Protestants (and especially the more radical Puritans). After all, the latter were the most vehemently opposed to Roman Catholicism and everything it stood for (and that meant opposing Catholic Spain). As a consequence, London and the Anglican Church went out of their way to avoid offending them. Religious liberty was the order of the day.

Despite this spirit of tolerance, many in the Puritan movement were concerned because they saw very little difference between the Anglican Church and the Roman Catholic Church. While the latter was run from Rome and the former was at least an English church, both relied heavily on an established clergy and an elaborate system of sacraments and liturgy. This ran directly in the face of Puritan thought, which believed that the process of salvation was much simpler—that it was a free gift, a result of Christ's death on the cross. Christ's death and resurrection absolved man of original sin, and all man needed to do was to accept this gift.[8] Sacraments, hierarchy, and liturgy merely got in the way, and Puritans considered them irrelevant. If anything, these men wanted to move religion in an even more "radically Protestant direction."[9] The Puritans not only wished to limit the power of the bishops, they wanted to eliminate these clerics.

Although the Presbyterians shared many Calvinist theological tenets with the Puritans, they differed because of their strong belief in the importance of a church hierarchy—not the kind the Anglican Church advocated, but one that had been properly ordained and trained. Presbyterians feared that permitting anyone to preach, as the Puritans advocated, would lead to theological chaos and confusion. The only way to avoid such a situation was to ensure that ministers and/or preachers had the Presbyterian Church's imprimatur.

With the defeat of the Spanish, however, the royal government and the state church became much more assertive. Who cared what the Puritans—with their tendency toward religious anarchism—thought? In fact, they were a negative influence, both politically and religiously. Better that they be brought back into the religious—and political—fold. Religious tolerance could be dispensed with.

This effort to enforce religious discipline was particularly evident when William Laud served as the archbishop of Canterbury. Laud considered his efforts to extend the Anglican Church's power and authority a sacred duty. For example, he tried to increase tithe payments, which had declined in real value in the preceeding century. He also attempted to use Church courts to reassert Anglican Church authority. Not surprisingly, this created considerable resentment on the part of large segments of the populace.

> Their excommunications, their prohibition of labour on saint's days, their enforcement of tithe claims, their putting men on oath to incriminate themselves or their neighbors—all of these were increasingly out of tune with the wishes of the educated, propertied laity, who were also critical of ecclesiastical control of education and censorship.[10]

One key point of contention between Laud and local citizens was lectureships. In contrast to the emphasis that Catholics (and many Anglicans) placed on the sacramental and ceremonial elements in religion, Presbyterians and others were more interested in preaching. A good minister was one who did an outstanding job telling people about the word of God; his ability to celebrate sacraments was secondary and, to many of the more fundamentalistic Protestants, irrelevant. It should be remembered that as congregations became more sophisticated and more educated, they demanded a higher level of preaching than the local clergy (who were not well trained in the art of homiletics) was either able or prepared to provide. Furthermore, the local parish minister was not usually chosen by the congregation but was nominated and appointed by the church hierarchy. Not surprisingly, individuals who were most supportive of the Anglican Church, its liturgy, and doctrine tended to be selected.

Because the populace could not have the kind of clergy it wanted, many people often turned to part-time lecturers. This meant that a congregation might put aside a certain amount of money to bring in a lecturer from the outside—even if the local parish minister was opposed. Not surprisingly, the Anglican Church leadership was upset and wanted to stop this practice. How could it maintain any kind of ecclesiastical control over what the average Englishmen heard if it could not determine who would be a lecturer? After all, what was to stop such individuals from teaching heresy? And even worse, these sermons often had political overtones.

They were accused of "popularity," of preaching sedition. There were few towns in the 1630s which were not having a quarrel with their bishop over such lecturers. Puritans were often accused of making an especial drive to buy patronage rights and endow lectureships in Parliamentary Burroughs, in order to influence elections.[11]

For his part, Laud carried on a constant campaign against lecturers and was able to suppress a large number of them. Laud's actions not only offended Puritans, they also alienated a number of the richer members of congregations, who wanted their own preachers and were prepared to pay for them. Why should this Anglican clergyman be able to stop congregations from spending their money? What if Laud and his friends considered these preachers to be disseminating heresy? What business was it of his? Congregations felt that they had a God-given right to practice their own religion, regardless of what the king and the Anglican Church might think about it.

Laud's attempt to recover money he claimed was owed to the Church also upset many of the gentry. A number of them—including Oliver Cromwell—owed their present status in life to properties that had been seized from the Catholic Church during the break with Rome. This was a direct threat to their property rights. What would Laud's efforts to gain back Church property mean for someone who owed his wealth to the buildings and lands that he had obtained from a former monastery? Should he be forced to pay taxes on them? Or would he be forced to give them back? Many of these men were members of Parliament, and they were not about to permit either the king or Laud to do anything that might undermine their economic and political status.

In a religious sense, Laud and Charles I seemed to many Protestants to be abandoning the main tenets of the Protestant Reformation. As the leader of the Anglican Church, Laud appeared to favor rituals that smacked of Catholicism, and he and Charles I seemed to favor Catholics at court and in determining high-level assignments in the government and in the army. In fact, given Laud's tendency to surround himself with Roman Catholics, many people began to believe that he was secretly a papist—an idea that was anathema to the Independents or evangelicals.[12]

As coming years would tell, Laud and Charles I were up against some formidable opponents. There were the Presbyterians, strongly entrenched not only in

Scotland but in the English Parliament as well. Bitterly anti-Catholic, they could not stand the idea of a return to a formal, state, "Catholic" church (even if in the guise of an Anglican institution). Needless to say, from a religious standpoint, the evangelical groups such as the Puritans were even more strongly opposed to Laud's efforts to re-centralize control over religion.

Parliament was the center of opposition to efforts by both the Church and the king to reassert London's control over the rest of the country. While some Parliamentarians were upset on religious grounds, others saw Charles's efforts for exactly what they were: an attempt to undermine if not eliminate Parliament's authority vis-à-vis the king. Many feared that if something was not done to stop Charles, he would soon have unlimited taxation powers. And this would be only the tip of the iceberg. Other liberties enjoyed by Englishmen would also soon be gone. In fact, many people feared that before long, Charles I would be able to dictate not only their religious beliefs but the rest of their lives as well.

Oliver Cromwell and the English Civil War

Like Robespierre in the French Revolution and Vladimir Lenin in the Russian civil war, Oliver Cromwell soon emerged as the most important personality during the events of the 1640s. Others such as Sir Thomas Fairfax, commander in chief of Parliament's army, would play a crucial role, but when it came to the person who left an indelible mark on this period, Cromwell was clearly the key player in both a political and a religious sense.

Oliver Cromwell was born on April 25, 1599, in Huntingdon. The family's wealth was based on his grandfather's acquisition of Church lands, and the revival of his own economic fortunes was also tied to "the acquisition of preferential leases on cathedral properties."[13] Indeed, part of Cromwell's obsessive anti-Catholicism might have resulted from a fear that his title to his land could be open to question if a pro-Catholic king succeeded in unmaking the English Reformation.

In contrast to many who would play an important role during the civil war, Cromwell was neither a full-fledged member of the gentry nor a commoner. In fact, as his biographers have noted, he tended to float between the two worlds. At one point he would play up his semiaristocratic background, while at another time he would stress his ability to understand and interact with individuals from the lower classes—as one of them, he would argue that he understood their problems. In fact, he would later make good use of this ability to relate to average people when he became commander of the New Model Army.

> Cromwell was not the typical country squire; the secure, obscure gentleman who rose from solid respectability to govern England with all the experience and all the limitations of a godly magistrate. His economic and social standing was far more brittle

than that implies; his reference to himself as being "by birth a gentleman, having in neither any considerable height, nor yet obscurity" takes on a tenser, more anxious patina.[14]

This mixed background was to have a profound impact on Cromwell when he became a soldier. Having lived among both common people and the gentry, he would turn out to be one of the few generals who was able to deal successfully with individuals from both worlds.

During his youth, Cromwell attended a local high school where his headmaster was Dr. Thomas Beard. Beard was a Puritan, who wrote tracts that were bitterly anti-Catholic. In addition, he often argued that the English Reformation was incomplete—and that contrary to Laud's efforts to restore the old church liturgy and structure, these should move further in a more radical and less structured direction. To Beard, the pope was the Anti-Christ. In one of his tracts he argued that "the whole of existence is a struggle between God and the powers of darkness, in which the elect fight for God and are certain of victory in so far as they obey his laws."[15] Furthermore, Beard maintained elsewhere that princes were not exempt from punishment because of their sins, and he noted that private property was sacred and that not even kings could seize it.

When it came time for college, Cromwell went to Sidney Sussex College, a newly founded Puritan institution. Some called Sussex College a "hotbed of Puritanism," and, as Christopher Hill noted, the college managed to avoid having its chapel consecrated by the Anglican clergy in spite of Laud's efforts to the contrary.[16] Oliver spent only a year at Sussex before he was forced to return home when his father died.

For our purposes, the most important legacy of both Beard and Sussex was that they seem to have imparted a deep sense of anti-Catholicism to young Oliver. Indeed, if one constant thread ran throughout his life, it was a deeply felt—and almost irrational—hatred of Roman Catholicism and those who adhered to its tenets.

On August 22, 1622, Oliver married the daughter of a well-off London businessman. His marriage was successful and his contacts with the rich and famous began to grow. He soon knew everyone who counted in England—many, if not most, on a first-name basis. These contacts would serve him well in later years. Of particular importance would be his close connections with people who were in the process of organizing opposition to the king in the Parliament.

Although the exact date when it occurred remains unclear, by 1635 Cromwell had undergone a spiritual conversion. As far as the nature of the conversion is concerned, it appears that he moved in the direction of the Independents (as the Puritans were often called), in believing that he had a personal relationship with God and that most ecclesiastical structures were both unnecessary and irrelevant. In any case, he clearly appears to have adopted the Independent/Presbyterian concept of predestination. "Cromwell manifests the Calvinist belief of a ceaselessly active God, a God of battles who in his overarching cosmic strategy as well as his attention to the smallest details, helps those who are on his side."[17] And throughout his political career, Cromwell never doubted that God was on his side.

In discussing Cromwell's attitude toward religion, we must understand that with the exception of Catholics, Cromwell did not show a marked preference for one religious orientation over another—at least, not during the early years of the Civil War. In time, he would oppose efforts by Presbyterians to force their religious orientation on the rest of the country, but for practical purposes, he seems to have been relatively neutral. He personally favored the more Independent approach, but he was smart enough to understand that if he wanted people to work together and, more important, to fight together, he needed to be tolerant. Indeed, whether for personal or political reasons, he consistently pushed religious tolerance—unless, of course, Catholics and especially Irish Catholics were involved. Like most other Englishmen of his time, he considered the Irish to be culturally and intellectually inferior to the English and to be under an evil religious spell. As a consequence, he would have found it hard to understand how anyone could argue in favor of tolerance for either the Irish or Roman Catholics. They were evil and, especially in the former instance, inferior from every point of view.

In 1640 Charles I decided it was time to call a new Parliament into session. Cromwell was a member of the existing Parliament. The problem was that Parliament had insisted that Charles make peace with Scotland, but he refused. Faced with this opposition to his plans, he dissolved Parliament. Meanwhile, lacking public support for his fight against Scotland, Charles was forced to conclude peace in October 1640. The terms of this peace compelled him to call another session of Parliament (the so-called Long Parliament, which was to last eleven years) into session. Again, Cromwell was a member of Parliament.

The new Parliament took a number of steps to counter Charles's efforts to increase his power. Parliament began by abolishing the courts that Charles relied upon to collect taxes. Indeed, any taxation without Parliament's approval was declared illegal. Parliament ordered that Archbishop Laud be imprisoned, and it impeached other ministers and judges. Bishops were excluded from the House of Lords, and Parliament declared that it could not be dissolved without its own consent. At the same time, many of the country's churches reversed the actions that Archbishop Laud had taken to make them more "Anglican."

Relations between Parliament and the king worsened. Finally, in October 1641 a revolt erupted in Ireland, which quickly led to military clashes. The immediate question was: Who would command the army that would be sent to Ireland to suppress the revolt? Parliamentary opponents attempted to use this opportunity to force the king to dismiss certain of his councilors and accept ones that had been approved by Parliament in their place. In the face of even more blatant attacks on the crown, swords were drawn in the House of Commons on the night of 22 November. Six weeks later the king tried to arrest the leaders of the Parliament. They fled to London, while the king went to the highlands of Scotland to seek support. War between the two sides finally broke out in August 1642.

The problem facing Parliament was how to raise an army. From a sociological standpoint they were in a bind. The Scottish army, with its strong Presbyterian orientation, could be counted on to come to Parliament's defense in a crunch. But when it came to building their own army, it was clear to Parliamentarians that their primary source of support was the lower classes. The problem, however, was that despite their opposition to centralized control by the king, the men from Parliament were far from egalitarians. Parliamentarians feared that if radicals, merchants, artisans, and similar folk played too great a part, these groups might try to increase their role in politics once a peace treaty was signed. Although this fear of the lower classes, and especially of the radicals, would continue to be a concern among some in Parliament for years to come, the fact was that Parliament needed an army *immediately,* and if that meant relying heavily on the lower classes, so be it.

The New Model Army

The situation facing Parliament's leaders was more difficult than the one that would face the Convention during the French Revolution or Lenin and the Bolsheviks during the Russian Revolution. In both latter instances, those who were fighting the old system were able to rely on large segments of the preexisting military. Their task was not to create a military out of nothing but to control people whose value systems were still tied to the old regime.

The English Parliament, however, had to create a totally new military—one based on very different principles. Fortunately, Parliament was able to tap into officers who had been soldiers in a wide variety of armies. However, as one observer noted, "It was not trained officers who were wanting . . . but trained soldiers."[18]

Parliament's first attempt to raise a military took place in 1642, as the fledgling army was placed under the command of the earl of Essex. It was poorly organized and while it included some twenty regiments of infantry as well as cavalry and dragoons, it was not effective and "rapidly crumbled away."[19] On the other hand, in the summer of 1644, with the help of Scottish troops, Parliament defeated the king's forces at Marston Moor. However, the generals were unable to follow up this victory, and within a few months the army had disintegrated. "This dismal record of failure had its natural consequence: discord and discontent within the armies, in parliament, and in London, as people cast about for someone to blame."[20]

It should be noted that in order to obtain the assistance of Scottish troops, Parliament had previously signed the Solemn League and Covenant.[21] This document stated that once the war had been won, Parliament would establish Presbyterianism as the country's state religion. What was important about this document was that like so many other things, it was agreed to under pressure and would later lead to a lot of dissension on the part of Parliament's supporters—especially its more radical members. Some of the more pro-Presbyterian members would cite it when the time came to appoint officers and chaplains, while the Independents tended to ignore it. In one instance, for example, a lieutenant was suspended by a Scottish major-general because he "had expounded Baptist views."[22] He was reinstated, but the incident led to dissension among senior officers who objected to the Scottish general trying to force his religious views on them.

Meanwhile, Parliament was fully aware of the need to come up with a better organizational structure for the military. On January 9, 1645, a plan for the New Model Army was submitted to Parliament. After much debate, the law establishing the army, with Sir Thomas Fairfax as its commander, was adopted two days later. In obtaining the House of Lords' and the House of Commons' approval, two conditions were agreed upon. First, both houses had to agree to the appointment of all officers. Second, all members of the military would have to accept the Solemn Covenant signed with Scotland, "but it was successfully argued that men could not be expected to adhere to a church government not yet in existence."[23] In theory, this amendment meant that no one could be commissioned in the New Model Army unless he accepted orthodox Presbyterianism. (Soldiers were exempted because it was feared that many would use their refusal to accept Presbyterianism as a means to avoid military service.) In practice, however, this latter condition requiring officers to be Presbyterians was a loophole that would be used repeatedly to keep individuals in service, regardless of their religious affiliation (with the exception of Roman Catholics, whose assistance was not wanted). According to one source, the Presbyterian-oriented House of Lords objected to almost 30 percent of the officers whom Fairfax proposed.[24]

Parliament's plan called for creating a new army of 6,000 cavalry, 1,000 dragoons, and 14,400 infantry.[25] In obtaining personnel, the New Model Army drew on the armies of Essex, Manchester, and Waller. In fact, all three armies (which had been fighting on behalf of Parliament) were under strength as a result of the campaigns of 1644. As a result, they were able to supply less than half of the infantry required by the New Model Army. Since there were not enough volunteers, it was necessary to rely on impressment. "More than half the infantry of the New Model were therefore pressed men, and yet, when the army took the field in May 1645, it was still 3,000 or 4,000 below its proper numbers."[26]

According to the original documents creating the New Model Army, Oliver Cromwell was not part of it. The reason was simple: In an effort to keep its more radical members from joining the armed forces, Parliament had passed a law, called the Self-Denying Ordinance, which prohibited members of Parliament from serving in the military. The problem, however, was that Cromwell was one of the country's best and, up to that point, most successful military officers. He was one of the few senior military officers (he had been fighting as head of an outfit called "Ironsides," which he had personally raised and which was the largest unit taken into the new army) who took the initiative, and, as such, he had won great respect from many of his colleagues. At the same time, he was also viewed by more conservative members of the government as "one of those 'violent spirits' given to 'agitation' and asperity' who so alarmed" many of the more conservative members of Parliament.[27]

In fact, Cromwell had been serving in the army for the past two years. His commission originally came from Parliament and in April 1645 the House of Commons extended his command. The House of Lords went along, but a number of more conservative members had spoken against it, arguing that it was a breach of the Self-Denying Ordinance. In the meantime, the position of the head of cavalry, the lieutenant general of cavalry, had been left vacant by Fairfax, who wanted Cromwell, and only him, to fill it.

On May 31, parliamentary forces suffered a defeat at Leicester. This unexpected and sudden reverse led to panic in the capital. Fairfax wrote to both houses, requesting Cromwell's appointment—some six days prior to the critical battle of Naseby.

> At Stony Stratford on 8 June, with a major battle clearly imminent, he put it to a council or war that parliament should be urgently requested to appoint Cromwell to the vacancy. His officers warmly and unanimously agreed; the Commons voted the appointment at once, but the Lords still smarting under the Self-Denying Ordinance's affront to their order did not. Fairfax did not wait, and when Cromwell in response to his summons rode into the army's quarters at 6 a.m. on the 13th he was welcomed "with a mighty shout."[28]

In a rather unprecedented move, the House of Commons decided to ignore the requirement that the House of Lords pass on such appointments. Faced with this situation, the House of Lords agreed to another three-month extension. In the end, the overwhelming victory at Naseby on 14 June would serve to override the House of Lord's objections and Cromwell would "secure for him the commission that by now he fiercely coveted."[29] Cromwell's political and military career was in full swing.

Before turning to a discussion of chaplains, it is worth looking at Cromwell the military commander since he had such a major influence both on the individuals appointed as chaplains and commanders and on the gospel that chaplains taught his troops.

Cromwell was unique among even parliamentary commanders in his understanding of, and ability to relate to, the average citizen. In the 1630s, for example, when Charles I tried to raise money by fining those who encroached on the royal forests in order to cultivate the area, Cromwell became their defender. In the event, "he made himself the spokesman of humbler and less articulate persons."[30] Unlike many from the gentry—who opposed Charles I for economic reasons— Cromwell was much more sympathetic to the demands of the lower classes. He was not as put off by them as were many of his colleagues, and having lived among them, he understood their fears, concerns, and dreams far better than others in Parliament—who wanted to keep the country's class structure, while either getting rid of the king or at least stripping him of his power over their economic activities. As Christopher Hill noted, "Cromwell was one of the few who had no inhibitions about using the loyalty and enthusiasm of the lower-class radicals."[31]

When it came to religion, there is little doubt that Cromwell favored the Independents since he counted himself as one. In fact, however, with the exception of Roman Catholics, Cromwell was very tolerant of the religious beliefs of those who served in the New Model Army. As he noted, after the very difficult battle of Bristol in September 1645,

> Presbyterians, independents, all had here the same spirit of faith and prayer: the same presence and answer; they all agree here, know no names of difference: it is that it should be otherwise anywhere.[32]

In another instance, Lieutenant General Cromwell got into a quarrel with Scottish Major-General Crawford, who was a professional soldier. Crawford had cashiered the officer mentioned previously because the man refused to sign the pledge of loyalty to orthodox Presbyterianism. He accused the officer of being a heretic.

> "Are you sure of that?" Cromwell retorted. "Admit he be, shall that render him incapable to serve the public?" Cromwell then laid down his own principle, the modern principle which then appeared terribly revolutionary. "Sir, the state in choosing men to serve them takes no notice of their opinions; they be willing to faithfully serve them, that satisfies."[33]

Within limits, for Cromwell, what a man did on the battlefield was far more important than the orthodoxy of his religious ideas. To quote Jonathan Adelman,

> In the more modern New Model Army, Oliver Cromwell replaced the standards of birth and privilege with efficiency and merit in recruitment and promotion of officers. By the late 1640s the commanders from the upper gentry (such as Essex, Manchester and Waller) were gone, and those few that remained (such as Fairfax) were genuinely talented. The key to success was the recruitment of talented officers from all classes and religious persuasions, not simply from a thin noble stratum, and the open rejection of professional mercenaries.[34]

Even though he tolerated those with different religious backgrounds, Cromwell's own belief in the importance of what he was doing was absolute. He was on a mission, following God's will.

> There was never any doubt in Cromwell's mind that he understood perfectly the mind and the plans of God. Trumpeting the crushing victory at Naseby, he told the speaker, "Sir, this is none other but the hand of God; and to him alone belongs the glory." A month later, after the equally decisive victory at Langport, he demanded of a member of parliament, "to see this, is to see the face of God!" The victory at Bristol two months after that was, beyond the shadow of a doubt, "none other than the work of God. He must be a very atheist that doth not acknowledge it."[35]

In this sense, Cromwell was not unique among Parliament's senior officers. Take, for example, Lord Fairfax. When his armies besieged Bristol in the summer of 1645, the whole region was infected with the plague. His officers advised against attacking the city, "but Fairfax brought them around to his view, declaring, 'as for sickness, let us trust God with the army, who will be as read to protect us, in the siege from infection as in the field from the bullet.'"[36] In another instance, Major General of Infantry Philip Skippon exhorted his troops with the words "Come, honest brave boys, pray heartily and fight heartily and God will bless us."[37] Or as Adelman observed,

> The men of the New Model Army saw themselves as free men consciously motivated by their beliefs rather than professional soldiers driven by mercenary and coercive considerations. Increasingly, over time the men and officers were deeply moved by both fundamentally political and religious beliefs, often inextricably linked.[38]

Turning to the issue of leadership, Cromwell's ability to inspire his troops was almost unmatched at this time. As one analyst put it,

> One of Cromwell's greatest strengths was his power to communicate to his officers and men his own utter certainty that they were instruments of a divine plan in which England had a special part to play, as an elect nation; they were the shock troops of the people of God. He upheld their right to worship and pray—and to preach too, if they were so moved—as their conscience directs them.[39]

In assessing Cromwell's contributions in the military sphere, it is important to note that he was not a professional soldier. Until he took up arms in defense of Parliament, he had almost no military experience at all.

> The most remarkable fact about Cromwell's military career is that he was forty-three years old and nearly three-quarters of the way through his life before it began. He had no experience of soldiering in 1642—not even of the recent Bishop's wars, no doubt to his relief.[40]

What Cromwell had, however, was an innate ability to train, discipline, and motivate soldiers. When it came to discipline, for example, Cromwell's troops were known as the best disciplined in the entire army. They were also some of the best trained in the entire military. Indeed, his accomplishments in this latter area were most notable. All of his troops had to learn their new trade in a very short amount of time. "That being so, their strength lay in their discipline, their commitment and their total mutual confidence, rather than in exceptional skills in manoeuvre or horsemanship."[41] His own commitment to the cause for which he fought was so strong and his ability to communicate with his soldiers so outstanding that he inspired those around him to accomplish far more than normally might have been expected of them. They were engaged in a holy crusade, one that their God had especially ordained. As a result, they were expected to give more than their best, and the record indicates that they did just that.

To recapitulate: Cromwell was a very religious individual, absolutely convinced of the righteousness of his cause. However, he also understood the importance of religious tolerance when putting together an army that was based in part on religious fervor. He related well to his officers and men, regardless of their class background. Indeed, there were instances when he made commoners officers—an action that raised a lot of eyebrows.

Perhaps because of his religious background and his recognition of the diversity of beliefs in the New Model Army, Cromwell understood the importance of having men who could rally these troops and help get them to fight as one, despite whatever theological differences they might have had. It was for this reason that he and Fairfax relied so heavily on chaplains. Chaplains would play a crucial role in ensuring not only that the army fought as one but that it emerged victorious.

Chaplains in the New Model Army

The duties of a chaplain in the Parliamentary Army were not explicitly laid down. Parliament decided that "the Assembly of Divines was asked to supply Fairfax with the names of 'godly and learned ministers' whom he could approach to staff the army."[42] In effect, chaplains would motivate troops, counsel them, ensure that morale was high, and provide political socialization, in the sense that they would constantly explain to soldiers how they were carrying out God's will.

Concerning the actual appointment of chaplains, the primary person involved was the unit's commander—a colonel. Prior to the New Model Army, colonels tended to appoint ministers whom they already knew as chaplains, men whose preaching they liked or whom they got along with well personally. For this reason, when the colonel left the unit, the chaplain usually did as well.

In some cases, the chaplain might be a Presbyterian, while in other instances, he might be an Independent.[43] This is why most chaplains in the duke of Essex's army were Presbyterians. They had been chosen because of their outspoken opposition to Laud's reforms. In addition, the overwhelming majority held ecclesiastical appointments.

Service as a chaplain was a temporary assignment. Most chaplains in the army were there for a year or less. There were few Independents in the army, primarily because very few soldiers would admit that they were Independents during the early years of the war.

As the conflict went on, a certain degree of religious radicalism among members of the army led to a need for greater numbers of Independent chaplains, and in fact, when comparing the number of chaplains between 1642 and 1645, the proportion of Independents increased somewhat. This is not to suggest that colonels necessarily appointed chaplains whose beliefs were closest to those of the rank and file. The primary concern was to get chaplains who would follow the wishes of the command and be able to relate to the average soldier on a religious basis. In the main, however, most chaplains had far more in common—from both a social and a religious standpoint—with the military leadership than they did with the soldiers they served.

By the time the New Model Army was created in 1645, the process for becoming a chaplain was clear. He was appointed by a regimental colonel and then commissioned by the commanding general. Yet there was no uniformity in the process of ordination. At least one-third had received Episcopal ordination in the Anglican Church, though some had renounced it. Others had been ordained as Presbyterian ministers. Still others had not been ordained but "had been called by a gathered congregation."[44] In accordance with Cromwell's attitude toward officers in general, formal ordination was far less important than the individual's ability to get the job done.

By almost any standards—but especially for clergy—the pay of chaplains was good. Upon joining the army, the chaplain received 20 pounds to enable him to purchase the necessary equipment. In addition, he received 8 shillings a day to help defray the costs for food and lodging. His yearly salary was 146 pounds.[45]

One major difference between the role played by a chaplain in the New Model Army and that which was to follow for the commissar in the French and Russian cases was that it was not the chaplain's task to convert the soldier to any particular form of religion.

> The Earl of Essex's Laws and Ordinance of Warre of 1642, which was the only disciplinary code for the parliamentary forces during the civil war, says little. It contains injunctions against blasphemy, unlawful oaths, and scandalous acts and lays down both penalties for absence from sermons and public prayers, but nothing about chaplains themselves.[46]

In practice, the chaplain had a number of jobs. His primary task was to provide spiritual assistance and in the process persuade the soldier that he should be prepared to go to war against his fellow countrymen and if necessary give up his life in pursuit of Parliament's cause—a primitive form of political socialization. This goal could be accomplished in a number of ways. For example, chaplains gave sermons in which they outlined clearly why the war was just and why a soldier should be prepared to do battle.

> For example, before the battle of Dartmouth on 19 January 1646, the weekly newsheet *The Moderate Intelligencer* (No. 47) observed that "Mr. Peters, and Mr. Dell, the most eminent Orators and Ecclessticks, who influence uses to be authoritative upon the soldiers, made learned and affectionate Orations to them, assuring all such as shewed valour in that action . . . great reward."[47]

In addition, chaplains also relied upon what the English called "fasting" and "humiliation."

A day of humiliation was held . . . to discover if the Lord wanted the army to advance on the enemy in Devon. Days of humiliation and fasting continued to be a favoured activity of the high command long after the emotional high points of the first civil war were behind them.

In addition to providing soldiers with spiritual advice and inspiring them to carry out their duties, chaplains also rebuked those who did not do so.

As for them of each nation who went away, they have by their ministers and others been so sharply reproved, and their fault in such sort aggravated that there is hope they will regain their credit by good service upon the next occasion.[48]

In this sense, the tasks carried out by parliamentary chaplains went far beyond what is expected of American chaplains today. While American chaplains may be required to help deal with things like combat fatigue, and while a few of them may have taken it upon themselves from time to time to chastise soldiers who did not fight efficiently, this was not a normal part of their job description, as it was for parliamentary chaplains.

Chaplains were also official spokesmen for the Parliamentary Army when it came to religious affairs. Even more surprising—given the American experience with chaplains—was the tendency to use them as messengers. For example, because of their educational background, they were often ordered to deliver concise reports "in person" to the House of Commons on the situation in the army. Such reports were normally given in the immediate aftermath of a major battle.

Both Fairfax and Cromwell employed the chaplains as news emissaries . . . to keep Parliament informed of the prisoners, ordnance, provisions, and towns taken in battle, to disclose the content of intercepted letters, and to present to the House gifts and trophies seized as prizes of war.[49]

Chaplains also sometimes acted as confidential agents for their commanders. Parliament often used chaplains as a means for carrying messages back to commanders. In short, chaplains appear to have been trusted both by their commanders and by members of Parliament as honest and competent message carriers.

Chaplains also played an important role after the battle. In particular, they were given the task of dealing with the civilian population. "For instance, after the battle of Torrington in Devon on mid-February, 1646, Peters preached to a large audience from a balcony in the market place because the church had been blown up."[50]

Another task that chaplains performed was that of war correspondent—in fact, they may have been some of the first war correspondents. It was up to the unit's chaplains to draw up "proceedings of the armies to which they were attached for publication in the press."[51] In this sense, they were official spokesmen both vis-à-vis the press and when dealing with the country's politicians.

However, by far the greatest and most important role played by chaplains in the Parliamentary Army came in battle. Their first task was to motivate soldiers, something they did several days in advance of an upcoming battle.

Naturally, the chaplains who accompanied the army on its marches played a key part in stirring up the fervour of the rank and file. They encouraged the conviction that the events the soldiers had witnessed were holy history, part of God's own plan for their country. Appointed by the army commanders, chaplains such as Hugh Peters and William Dell were powerful personalities in their own right, who did much more than simply echo the views of Fairfax and Cromwell. By their eloquence they greatly strengthened the conviction that the army was an almost passive instrument in the hands of the almighty.[52]

Concerning combat situations, we have a number of reports indicating that chaplains were in the midst of the battle, working to inspire their troops to victory. Here are a few examples:

The Battle of Edgehill

Chaplains appear to have been present all over the battlefield.

They rode up and down the army, through the thickest dangers, and in much personal hazard, most faithfully and courageously exhorting and encouraging the soldiers to fight valiantly, and not to fly, but now if ever to stand to it, and to fight for their religion, laws and Christian liberties.[53]

The Battle of Naseby

According to one source, Chaplain Hugh Peters "rode from rank to rank with a Bible in one hand and a pistol in the other exhorting the men to do their duty."[54]

The Battle of Bridgewater

Chaplains again played a key role in motivating troops.

Before the assault on Bridgewater[,] On the Lord's day, July 20, 1645, Mr. Peters in the forenoon preached a preparation sermon, to encourage the soldiers to go on; Mr. Bowles likewise did his part in the afternoon. After both sermons the drums beat, the army was drawn out into the field: the commanders of the forlorn hope, who were to begin the storm, and the soldiers being drawn together in the field, were there also afresh exhorted to do their duties (with undaunted courage and resolution) by Mr. Peters. . . .[55]

The Battle of Bristol

A day prior to the battle was declared a day of fasting. Then on August 29 a spiritual exercise was held, which was "directed at winning divine blessing upon the endeavour."[56]

The Battle of Oxford

On the day before the battle, a Mr. Saltmarsh preached before the headquarters contingent.

> Mr. Dell preached in the forenoon and Mr. Sedgwick in the afternoon; many soldiers were at each sermon, divers of them climbing up the trees to hear for it was in the orchard before his Excellency's tent, and it is very observable to consider the love and unity which is amongst the soldiers, Presbytery and Independency making no breach.[57]

The key question, however, is to what degree was this system of motivation and morale building successful? Based on the information that we have today, it appears that the parliamentary chaplains did a good job of motivating the soldiers they served with. Christopher Hill notes, for example, that "from the beginning, Oliver's men kept their powder dry as well as trusting in God."[58] Or as another observer put it,

> The conviction that they belonged to an army bathed in the rays of providential favour, that they were fighting "the warfare of heaven," "overcoming evil by doing good," that God was exceedingly bountiful in his goodness towards them, had a liberating effect throughout the army which led to many acts of exceptional courage and improvisation.[59]

It was said that while in other militaries men became less civil, in the New Model Army "men grew more religious than in any other place in the kingdom."[60] Indeed, according to Solt, this was a viewed shared by royalists. "Two Royalists, William Chilingworth and Clarendon, found a great deal more piety and sobriety among the officers and soldiers of the Parliamentary Army than among the King's troops."[61] While it is impossible to determine retrospectively just how important chaplains and religious faith were in the English Civil War, those involved seem to have believed that it was crucial. In discussing the Battle of Bristol in 1646, for example, William Dell made the following observation: Bristol

> was conquered by faith more than by force; it was conquered in the hearts of the godly by faith, before ever they stretched forth a hand against it, and they went not so much to storm it as to take it, in the assurance of faith.[62]

This helped explain, according to the same author, why the New Model soldiers were able to defeat larger numbers of Royalists.

Christopher Hill put the role of religion in inspiring the soldiers of the Parliamentary Army in proper context when he observed,

> If for "God" we substitute some phrase as "historical development" or "the logic of events," as the Puritans almost did, then there can be no doubt that powerful imper-

sonal forces, beyond the control of any individual will, were working for Cromwell and his army. The evidence for this is not merely that Parliament won the civil war; it is also the complete inability of the old government rule in the old way which had been revealed in 1640, its financial and moral bankruptcy.[63]

In essence, these soldiers were cooperating with God. "God works through individuals, and the success of a virtuous human being is at once his victory and the victory of divine grace working in him."[64] The only difference between these soldiers and those who would support the French Revolution was that the latter had substituted history for the will of God.

In this context, it is important to reiterate something noted previously—chaplains did not work to shape the religious character of their New Model Army. That was not their task. Rather, they seized on the existing religious attitudes and used them to motivate the troops they served with.

The accomplishment of Cromwell's clergy is even more impressive when one considers how few chaplains served in the Parliamentary Army. Two chaplains were always stationed at headquarters, while about thirty others served in the different regiments. Indeed, in some cases regiments did not have chaplains, thereby making it necessary for certain chaplains to commute from one regiment to another, especially as the probability of combat increased. If we look at the entire period from 1645 to 1651, a total of forty-three chaplains served in the New Model Army. In terms of religious preference, no more than four were Presbyterians in 1645, "and this number had shrunk to zero prior to 1649."[65] As time went on, the strength of the Independents in the army grew—to the point where the more formal Presbyterians found themselves having greater problems communicating and relating to the rank and file.

One reason both for the decreasing number of Presbyterian chaplains and for the low numbers of chaplains required by the military was the presence of lay preachers. Most of these men were Independents—and some were radical Baptists.

Chaplains faced increasing competition from lay preaching, especially in the New Model Army. This seems to have set up a vicious cycle. There was a shortage of chaplains, so laymen began to preach and . . . the resulting spread of sectarianism discouraged ministers from joining the army as chaplains.[66]

Given the strong sentiment against lay preaching among Presbyterians in Parliament, the latter passed an order in 1645 prohibiting preaching by anyone who was not an ordained minister. The order was sent to Fairfax, with instructions to ensure that it was strictly observed and that those who violated it were severely punished. In fact, soldiers in the New Model Army paid little attention to it, and the officers were not about to go out of their way to enforce an order that would have undermined morale.

As time wore on, the degree of religiosity present in the military was less and less dependent on the presence of chaplains. The troops themselves became increasingly religious, ignored Parliament and the Presbyterians, and preached the gospel to each other. This is not to suggest that chaplains did not continue to play an important role. They did, especially when it came to motivating troops prior to battle, but the need for them on an everyday basis declined. For example, when Cromwell set sail for Ireland in 1649 to put down the rebellion in that country, it appears that he had only a few chaplains in his army. John Owen and Hugh Peters served as his own private chaplains. "Owen soon returned to become Dean of Christ Church and Vice Chancellor of Oxford." Peters did not stay much longer. Only Thomas Patient, who founded a small Independent congregation at Waterford, stayed for any length of time.[67]

Conclusion

By 1647 the First Civil War was at an end. Parliament had triumphed. Internal battles were not over—there would be a lengthy struggle between the New Model Army and Parliament and, eventually, Cromwell would emerge as the "Protector" or supreme ruler of England. In the end, the monarchy would also be restored, albeit never again as an absolute monarchy.

For our purposes, the "golden age" of chaplains ran from 1642 to 1647. As this chapter has pointed out, chaplains played an important role in Parliament's army. They dealt with a group of men for whom religion was very important. It was their task to utilize this sense of religious conviction not only to motivate these men but to mold them into an effective fighting unit and ensure that they were loyal and their morale was high. In this task chaplains were successful.

Just as we have noted what the parliamentary chaplains did, it is also important to note what they did not do. They were not used as watch dogs to ensure the political reliability of the line officers with whom they worked. In this sense, I strongly disagree with Jonathan Adelman, who on two separate occasions referred to chaplains as "just like commissars," or as "the seventeenth-century version of modern commissars."[68] Instead, chaplains were subordinate (in a military sense) to line officers and served at their pleasure.

As far as I can determine, chaplains did not have a major impact on military decisions. This is not to suggest that line officers might not have asked them for spiritual guidance and that this advice might not have influenced how and when they fought battles. On the other hand, there was no requirement that chaplains had to sign off on orders that the line officer issued—to ensure that the officer did not act against the interests of the regime, as happened in the Soviet case.

One major difference between the English Civil War and those events following it was that the overwhelming majority of officers and men shared the same value system as members of Parliament—at least, insofar as opposition to an absolute monarch and an established Anglican Church was concerned. As time went on and the influence of the Independents within the army grew, serious differences between Parliament and the New Model Army would arise, but in the war, at least, they were united in their opposition to Charles I. There was no need to make major modifications in values—the kind of problem that would face revolutionaries in both the French and Russian revolutions. Cromwell and Fairfax were able to rely on traditional religious values (those that arose out of the English Reformation) to inspire their troops.

Indeed, whatever else one may think of Cromwell, who was often the power behind Fairfax, he was an outstanding military leader. This is true regarding not only his tactical sense but also his ability to deal with his troops. He understood, better than many military leaders, how to handle soldiers from the lower classes. Similarly, he also knew the importance of religion and religious tolerance (with the exception of Roman Catholics) in leading men. Religion was paramount in soldiers' lives, and through the use of chaplains Cromwell was able to tap this indispensable source of inspiration and motivation. In short, chaplains were key and Cromwell knew it. Without their help, it is doubtful that Cromwell and his army would have enjoyed the success it did.

Notes

1. The discussion of the period prior to Cromwell is based on John Smyth, *In This Sign Conquer* (London: A. R. Mowbray, 1968), 1–16.

2. Smyth, *In This Sign Conquer,* 14.

3. Smyth, *In This Sign Conquer,* 17.

4. This was not the first time that chaplains were utilized outside of Britain in this manner, although it may have been the most systematic use of them in this regard to this point in time. Washburn has argued, for example, that "Probably all of the armies participating in the Thirty Years' War had their chaplains." "Gustavus Adolphus depended not only on a rapidly moving army, but on one that was as sound morally and spiritually as he could make it. . . . To attain his purposes, he appointed a chaplain-general and two chaplains in each regiment. Fabritius, the chaplain-general, was given full direction of the religious department and was responsible for the morale of the army." Henry Bradford Washburn, "The Army Chaplain," *Papers of the American Society of Church History,* Second Series, vol. 7 (1923), 12.

5. S. P. MacKenzie, *Revolutionary Armies in the Modern Era* (London: Routledge, 1997), 8, 13. Even though there were more chaplains in Cromwell's army that MacKenzie admits (see the present volume, p. 73), that is not the key issue. As this chapter will demonstrate, while Cromwell could have always used more, they played an important role in a variety of areas.

6. Christopher Hill's *God's Englishman: Oliver Cromwell and the English Revolution* (London: Weidenfeld and Nicholson, 1970), 58.

7. Of all the works I consulted for this chapter, Hill's *God's Englishman* was by far the most useful.

8. See Leo F. Solt, *Saints in Arms* (Stanford, Calif.: Stanford University Press, 1959), 29, for a fuller discussion of this issue.

9. Hill, *God's Englishman*, 22.
10. Hill, *God's Englishman*, 19.
11. Hill, *God's Englishman*, 48.
12. Laud would eventually be arrested, imprisoned in the Tower of London, and executed in 1645.
13. John Morrill, "The Making of Oliver Cromwell," in John Morrill, ed., *Oliver Cromwell and the English Revolution* (London and New York: Longman, 1990), 20.
14. Morrill in Morrill, *Oliver Cromwell*, 24.
15. Hill, *God's Englishman*, 39.
16. Hill, *God's Englishman*, 40.
17. Ian Gentles, *The New Model Army* (Oxford: Blackwell, 1992), 93.
18. C. H. Firth, *Cromwell's Army* (London: Methuen, 1962), 15.
19. Charles Firth, *The Regimental History of Cromwell's Army*, vol. 1 (Oxford: Clarendon, 1940), xv.
20. Gentles, *The New Model Army*, 3.
21. See Alexander Crawley Dow, *Ministers to the Soldiers of Scotland* (Edinburgh: Oliver and Boyd, 1962), 88.
22. Gentles, *The New Model Army*, 3.
23. Mark A. Kishlansky, *The Rise of the New Model Army* (Cambridge: Cambridge University Press, 1979), 40.
24. Gentles, *The New Model Army*, 23.
25. Gentles, *The New Model Army*, 10.
26. Firth, *Cromwell's Army*, 36.
27. Morrill in Morrill, *Oliver Cromwell*, 47.
28. Austin Woolrych, "Cromwell as a Soldier," in Morrill, *Oliver Cromwell*, 103.
29. Gentles, *The New Model Army*, 27.
30. Hill, *God's Englishman*, 49.
31. Hill, *God's Englishman*, 60.
32. As quoted in Woolrych, "Cromwell as a Soldier," 97.
33. Hill, *God's Englishman*, 68.
34. Jonathan R. Adelman, *Revolution, Armies and War: A Political History* (Boulder, Colo.: Riener, 1985), 23.
35. Gentles, *The New Model Army*, 93–94.
36. Gentles, *The New Model Army*, 92.
37. Gentles, *The New Model Army*, 92.
38. Adelman, *Revolutionary Armies and War*, 26.
39. Woolrych, "Cromwell as a Soldier," 97.
40. Woolrych, "Cromwell as a Soldier," 93.
41. Woolrych, "Cromwell as a Soldier," 105.
42. Gentles, *The New Model Army*, 32.
43. The best works on chaplains in the Parliamentary Army are Anne Laurence, *Parliamentary Army Chaplains, 1642–1651* (Suffolk: Boydell, 1990), and Firth, *Cromwell's Army*, 311–345.
44. Laurence, *Parliamentary Army Chaplains*, 22.
45. Laurence, *Parliamentary Army Chaplains*, 15.
46. Laurence, *Parliamentary Army Chaplains*, 6.
47. Solt, *Saints in Arms*, 10.

48. Firth, *Cromwell's Army,* 319.
49. Solt, *Saints in Arms,* 11.
50. Solt, *Saints in Arms,* 11.
51. Firth, *Cromwell's Army,* 326.
52. Gentles, *The New Model Army,* 94.
53. Gentles, *The New Model Army,* 96.
54. Gentles, *The New Model Army,* 96.
55. Firth, *Cromwell's Army,* 318.
56. Gentles, *The New Model Army,* 72.
57. Firth, *Cromwell's Army,* 320.
58. Hill, *God's Englishman,* 65.
59. Gentles, *The New Model Army,* 105.
60. Solt, *Saints in Arms,* 14.
61. Solt, *Saints in Arms,* 14.
62. Gentles, *The New Model Army,* 95.
63. Hill, *God's Englishman,* 230.
64. Hill, *God's Englishman,* 244.
65. Gentles, *The New Model Army,* 95.
66. Laurence, *Parliamentary Army Chaplains,* 6.
67. Firth, *Cromwell's Army,* 322.
68. Adelman, *Revolutionary Armies and War,* 26, 27.

3

Political Commissars in the French Revolution

Soldiers, We call you back to the rigorous discipline which alone can cause you to win and which spares your blood. Those who provoke the infantry to disband in the face of enemy cavalry, those who leave the line of battle before or during combat, or during retreats will be arrested immediately and punished by death.

—Louis St.-Just and Philippe Le Bas, Representatives on Mission to the Armée du Nord

THE FRENCH REVOLUTION MARKED a major turning point in military history. First, it witnessed the introduction of mass armies (far larger than in the English Civil War) in place of the mercenary forces common up to that time. Second, because of the revolutionary nature of events, the new regime was forced to deal with a substantial difference in values between itself and other parts of society. The involvement of almost all citizens in the military, together with the momentous societal changes that France was undergoing at the time, meant that the old forms of political control no longer sufficed. New mechanisms had to be developed in order to ensure stable, efficient civil-military relations.

The ten years of the French Revolution—from 1789 to 1799 (when Napoleon seized power)—were chaotic, tumultuous, and unpredictable. Under the Jacobins, terror was as rampant in the military as it was in the rest of French society. An officer could perform well one day, only to find himself facing the guillotine the next, because of a change in personalities or politics. Indeed, as far as the French military was concerned, the French Revolution was the most politicized period in French history. One not only had to ensure that one's behavior corresponded with the views of the prevailing political authorities, one also had to be sure that one was performing well in the military sphere.[1] Failure to win battles could just as easily lead to punishment as could incorrect political views. Career longevity was by no means assured.

Given the fundamental changes that were under way within the French polity, the new regime had to come up with a vehicle to ensure its control over the armed forces. Indeed, there were periods when central government control over the military seemed very much in doubt. To make certain that Paris remained in control, a number of techniques were employed. These measures, however, were as confusing, changing, and chaotic as French society itself. These were not the highly structured and carefully thought out measures that would be employed in the Soviet and Chinese revolutions. The latter would build on the French experience, but their approach would be much more systematic. Authorities in Paris would go from one extreme to the other—from sending the commander of a unit to the guillotine because his troops complained about how he treated them to forcing soldiers who refused to obey orders to suffer the same fate. Similarly, officers who were sent to jail at one point because of their class background or some presumed personal or political fault could and did find themselves recalled to service—often in a high and very important position—months or years later.

In short, the situation within the French armed forces was as dynamic and unpredictable as it was complex and confusing. The control measures that Paris utilized and the criteria of political correctness changed, based on who was in charge of the country at a particular point in time.

The French Military on the Eve of the Revolution

The prerevolutionary French armed forces belonged to the crown. Soldiers and officers took an oath to the king. They were his troops and were obligated to follow and obey his orders. Indeed, one interesting aspect of the royal army (wearing white uniforms, the color of the House of Bourbon) was that in addition to the traditional obligation of defending the country against external enemies, the royal army was also obligated to protect the king against internal threats; in other words, the military was also responsible for helping to maintain internal order.

By modern standards, the French military was relatively small. In 1789, when the revolution began, the military stood at about 150,000 troops. It was divided into 112,000 infantry, 33,000 cavalry, and 6,400 artillery—numbers that would be dwarfed later not only by Napoleon's forces, but by those of the various revolutionary governments that were to follow the collapse of the old regime.[2] For recruits, the regime relied primarily on volunteers. Most of these individuals were men who found it difficult to secure employment and as a result turned to the military. One consequence was that the military enjoyed very little prestige in French society. Military service was even less popular among peasants. At first glance, this may seem surprising since 80 percent of the country was made up of peasants—but in 1789 only 15 percent of those in the army whose backgrounds could be identified were peasants. On the other hand, this low representation should not be too surprising, given the need for labor in the agricultural sector. From a political standpoint, the most important implication is that the peasant population had little sympathy and understanding for the military—a situation that would have serious consequences during the ten years of revolution. Most of the royal army's soldiers came from towns with populations of 2,000 or more and 20 percent of those came from towns with populations over 10,000. Furthermore, three-fifths of the soldiers had artisan backgrounds.[3]

Discipline was brutal. "Most officers . . . had taken a leaf from the Prussian book and wished to transform the soldier into an automaton." Anyone who failed to obey an order instantly or who got out of line could be forced to run a gauntlet while his colleagues hit him with rods or the flat of a sword. In most cases, officers were seen only when punishment was involved. As Bertaud put it,

> Poverty, humiliation, and contempt. The common soldier was scorned by his officers, and sometimes by the bourgeois, many of whom shut their doors and fastened their shutters on hearing of the approach of the military.[4]

In addition to its small size, the royal army also made use of mercenary units—loaned to the king by foreign princes or governments for payment. This was especially true of Swiss and German troops, both of which would find themselves in the middle of the French Revolution. The former, for example, would fire on a Paris crowd on August 10, 1792.

When the revolution began, there were a total of 102 battalions in the French Army. Of that number, 79 were French "and the others drawn from among mercenaries of Europe, from the Swiss cantons and the various German states, and from Ireland and Liege."[5] The problem with these units—once the revolution began—was that they had little personal loyalty to France or the revolution. They soldiered for money, and their primary political loyalty was to the region they came from.

Concerning the officer corps, the royalist army was manned almost entirely by nobles. In fact, 90 percent of all officers were from the nobility. In 1789, of 10,000 officers in the French Army, only 1,000 were commoners. And those commoners were almost always lower-ranking officers. For example, at the end of the old regime, "of 11 marshals, 5 were dukes, 4 were marquises, 1 a prince, and 1 a count. Of 196 lieutenant generals only 9 were untitled nobles."[6] As Scott put it, "An officer of common origins was always something less than a complete officer, no matter what his skills, no matter what position he held."[7] Needless to say, the monopolization of the officer corps by the nobility created dissatisfaction on the part of commoners—especially those who wanted to be officers or who were already officers and found their path to advancement blocked because of an accident of birth.

Becoming an officer—especially a senior one—was an expensive proposition. The ranks of captain and colonel could cost between 6,000 and 14,000 livres, while command of an infantry regiment might cost from 25,000 to 75,000 livres. Other officer grades were ostensibly open to all and did not have to be purchased. In practice, however, these positions were given to friends of the regimental commanders—for a fee, of course. Officers also were expected to maintain an active social life—a very expensive undertaking that required the use of private funds.

Enlisted personnel in the army on the eve of the revolution were made up almost exclusively of commoners. They were not well thought of by the civilian population and were alienated from peasants. Splits between officers and enlisted personnel were clear, because the former looked down on the latter. Even within the officer class, important differences existed—between what were referred to as soldiers of fortune (those commoners who had managed to achieve an officer's commission) and the rest of the officer corps. This is the main reason why the majority of non-noble officers and NCOs supported the new regime when the revolution occurred. As they saw it, the new regime offered them the possibility of rising to whatever rank their skills might permit. No longer would certain positions be closed to them because of an accident of birth.

At the same time, it is important to keep in mind that military skills were monopolized by the nobility. It would be hard—if not impossible—to put an army in the field or a navy to sea without making use of the nobility, a problem that was to haunt the new government. Regardless of what the political loyalties of nobles might be, their military skills were indispensable.

Although there were exceptions, the nobility supported the monarchy, especially once it started to come under attack. Meanwhile, discipline was maintained by draconian methods. As a result of the nobility's monopoly of the officer corps and support for the crown, as well as the strict discipline that reigned in the French Army, Paris did not believe it would be necessary to introduce special control measures when social disturbances broke out.

The Revolution and the End of the House of Bourbon

As the year 1789 began, the country faced a dismal economic future. The harvest in 1788 had been bad, and since the one from 1789 was not yet in, revolts were occurring all over the country. "Poor peasants and famished artisans besieged the markets, demanding grain at low prices."[8] Because the French system called for the army to become involved in the maintenance of internal order, troops were utilized to protect grain. Troops guarded warehouses and the shops of grain merchants. In addition, since grain had to be moved around the country, troops were also used to protect these convoys. In practice, this meant breaking up regiments into small units, which more often than not were commanded only by sergeants and corporals. Not only was discipline hard to maintain in the face of such long, forced marches and short food rations, the small units that troops now served in brought them into direct contact with the rebellious crowds who were demanding wheat. If anything, this latter experience helped undermine military cohesion, and serious questions were raised in the minds of many soldiers—whose relatives and family were often protesting—about the legitimacy of the royal government.

Despite their doubts, army troops put down a riot in Paris on April 27. As the situation deteriorated further, the royal government began to concentrate more troops around Paris. Indeed, by the beginning of July, some 17,000 troops were located in the area.[9] Conditions facing these troops were far from satisfactory. Food was scarce and almost nothing had been done to prepare for the troops' arrival. In the meantime, pamphlets encouraging soldiers to disobey their officers were distributed, and some troops began to fraternize with the local populace. For the first time, French soldiers—and not just in Paris—were being inundated with propaganda against the crown they served. Should they continue to support the king—to whom they had sworn an oath—in the face of efforts by the local populace to obtain the kind of equal rights that soldiers wanted for themselves in the military? Why should they use their weapons against people whose goals were the same as theirs—basic human rights? In the process of concentrating troops, the king and his advisers forgot that a basic rule of using the military for internal purposes was to keep troops isolated from the general populace to control their perception of reality. Since the government made no effort to isolate troops from the crowds, it was only a matter of time before soldiers began to have serious doubts about the tasks that might be assigned to them.

The impact of fraternization and the poor conditions in which the army was living became evident on July 14th, the day the Bastille was stormed, as five of the six battalions of French guards went over to the insurrectionists. Similarly, it soon became clear to the king's closest advisers that the military could not be used to close sessions of the newly created National Assembly, where the king's authority in a variety of areas was being questioned. Indeed, as one writer observed,

the refusal of obedience was itself a positive political action in the context of 1789, because if the army could throw its weight behind a political faction, it could also provoke a decisive change in the political balance by the withdrawal of its unquestioning support of the monarchy.[10]

Before we continue, something should be said about the legislature. Until June 17, the National Assembly had been only one part of a three-part legislative entity. The First Estate was made up of the nobility, while the Second Estate encompassed the clergy. The Third Estate was primarily made up of the bourgeoisie and well-off property owners, who claimed the right to speak for the rest of society. By June 17 it was becoming clear that the Third Estate encompassed all three estates and, as a consequence, the legislature's name was changed to the National Assembly.

Although the king's role was still uncertain at this point, the National Assembly had become the preeminent political institution in the country. What to do? How to make the National Assembly immune from royal interference?

To protect itself from the king—but more important, from the masses of poor people in Paris—the Assembly set up a National Guard. On July 31, the Military Committee of the Parisian Assembly of Electors recommended that no transients, domestics, workers, or artisans be permitted to join the guard. Toward this end, the Assembly stipulated that "all members had to pay for their own uniforms themselves. . . . Not only did it keep the lower orders out of the Guard, it also heightened the group solidarity of those who could afford to dress up."[11] As a further measure, the Assembly ordered that all citizens who were not members of the National Guard be disarmed. In fact, guard units had been established all over France. And unlike the Paris unit, which was under the control of the National Assembly, other units were under local control. As a consequence, their individual political orientation more often than not reflected the political preferences of the town or region in which they were located.

Meanwhile, the line army (the regular military) came increasingly under the influence of local societies or political clubs. Soldiers were invited to join local societies, where they often became radicalized. As far as the military structure itself was concerned, the most debilitating factor was the tendency of these groups—all of which were left wing—to support soldiers' complaints (whether valid or not) vis-à-vis their officers. "Wrongful imprisonment, savage punishments, careless phrases, or insulting rebukes were all deemed sufficient cause for the denunciation of an officer and the rejection of his authority."[12] There were notable exceptions—especially the foreign regiments, which could not communicate with either the societies or the crowds. In addition, some purely French regiments remained loyal to the central government. Everything depended on the attitude of the unit's NCOs and officers. If the latter could maintain discipline, whether by charisma or simple orders, the units remained intact and loyal to the king. Regarding officers and NCOs, they made it clear to the crown that while they would serve loyally, they believed that the only valid criteria for military position and promotion should be talent and merit. On paper, this meant that those of noble background were prepared to sacrifice their claims to rank and privilege. In reality, however, while promotion was now open to everyone, promotees were selected by the unit's officers—which meant a monopoly for nobles (since they made up the vast majority of officers, especially at senior ranks)—and officers made seniority the primary criteria for promotion.

Discipline continued to deteriorate. Indeed, Forrest had it right when he noted that

> In the aftermath of July 14, numerous individual acts of defiance occurred as soldiers refused to fire on the people, even when the king's authority was directly threatened. . . . The authority of officers was increasingly undermined as more and more soldiers placed their loyalty to the new political authorities above their established loyalty to the king and to their military experience.[13]

The number of desertions and mutinies moved steadily upward. In April 1790, for example, soldiers from different units fought a pitched battle for nine days because one unit considered the other to be loyal to the old regime. There were numerous other examples of disobedience, but the best known and most serious mutiny occurred at Nancy in August 1790.

Three regiments were stationed in that city, one French and two Swiss (one infantry and one cavalry). Relations between officers and men had been strained. The National Assembly had outlawed political associations and clubs in an effort to enforce discipline. The troops ignored the order and demanded an audit of regimental accounts. They then arrested the regimental quartermaster and confined all officers to their barracks. These developments quickly spread to the other units. Mutiny in Nancy was a fact of life.

Fearful of the loss of control over its military forces, the National Assembly ordered the Marquis de Bouille to restore order and strict discipline. Bouille set out to make an example of the mutineers and marched on the town with a large force of troops—many of whom were the more reliable foreign troops. Nancy was taken after three hours of fighting. Numerous soldiers were hung, many were sent to the galleys, and others were imprisoned.[14] While order was restored in the military at Nancy, the public reaction was one of outrage—noble officers massacring common soldiers. Despite the harsh manner in which this mutiny was put down, revolts of this kind continued into 1791.

Given the unsettled nature of things and the tendency in France and within the army to depict the nobility as the source of all that was wrong with the country, many of the country's officers began to ask themselves why they were continuing to serve. Some, like the Marquis de Lafayette, had fought in the American Revolution and supported the move toward equality, but others could not surmount their doubt and disillusion. Why should they be part of an army that seemed to despise them? After all, they could do nothing about their noble birth. If nonnoble birth was the primary criteria for appointing officers in the French Army, why should they remain in it? The situation was made worse by the fact that these noble officers had taken an oath to support the king. Meanwhile, the role of the king diminished (the end of feudalism was declared on August 11, 1789), and the revolution was increasingly radicalized. Growing numbers of such officers began to believe that there was no future for them in France. As a consequence, many of them began to emigrate.

The exodus of officers continued as attacks on the monarchy intensified. On October 5, the king was forcibly returned to Paris from Versailles and held under what amounted to house arrest. Then on June 21 the king and queen fled Paris, only to be captured a short time later and returned to the capital. In December the king stood trial and on January 16 was executed. As one writer put it, "By executing the king, they had severed France's last ties with her past, and made the rupture with the ancien régime complete."[15] Now, officers from the old regime had no reason to continue to serve in the army. How could their oath be binding if the king was no longer alive and if most senior members of the nobility had fled the country and were plotting to return at the head of a new monarchy? As a result, many officers joined their comrades in an attempt to reimpose the royal house in France from abroad. To quote Scott, "The turning point of the military emigration was the failure of the flight of the king on 21 June . . . by the end of 1791 there were perhaps as many as six thousand French officers on foreign soil."[16] As a consequence, it became increasingly difficult to find experienced officers—men who could assume command with some sense of confidence that they knew what they were doing. And this was as true of the navy as it was of the army. As one study of the navy during Revolutionary France pointed out, "In response to the dissolution of its authority, a large portion of the fleet's officer corps abandoned the service."[17]

In response, members of the National Assembly argued that a soldier owed obedience to the law and not to a particular officer. As long as the officer was carrying out the country's laws, his orders should be obeyed. In practice, however, this had little impact on day-to-day life in the French armed forces. More often than not, the officer's authority was questioned openly.

Not only officers were impacted by the slow deterioration of the army. Regular soldiers showed their displeasure with what was happening by deserting. For example, desertion in the line infantry was 1.87 percent in 1788. By 1789 it was up to 3.62 and then 4.88 percent in 1790.[18] Service in the French Army was becoming an increasingly unpleasant experience for all concerned.

Political clubs also continued to be a major problem for the French Army. These clubs were active in almost every garrison town and constantly undermined officers' authority. As soon as a unit arrived, the Jacobins invited the soldiers to attend Jacobin meetings and in the course of these meetings helped soldiers to understand their rights and defended them against any real or imagined violation of their rights by officers. Unfortunately, the primary impact of the Jacobin clubs was to further undercut order and discipline in the army. As Lynn noted, "Soldiers did consider a political club to be a forum in which to denounce their officers."[19] Indeed, reading through histories of this period, one comes away with the impression that the more radical Jacobin Clubs, in particular, devoted most of their time to undermining the military. "The Jacobin clubs both in Paris and in the provinces, held long sessions at which extreme anti-officer feeling was expressed. Officers were frequently assumed to be suspect by the very fact of their rank in the line."[20] These clubs had a negative impact not only on army units, but also on the navy.

> Port municipalities, often in alliance with Jacobin clubs, sought to undermine the governance of naval officers in keeping with the principle of Popular Sovereignty. They subverted the navy's efforts to enforce discipline over its sailors and workers, and they defended, even encouraged, the rising tide of mutiny.[21]

For its part, the National Assembly—which was the county's supreme executive organ—was very concerned about the constant deterioration in military order. As a result, in September 1790, it passed a law prohibiting soldiers from joining political clubs. The goal was to depoliticize the soldiers and to restore discipline.

Then on September 19 the Assembly banned contacts between troops and civilian societies in general.

In fact, the clubs did not go away, and under considerable pressure, the National Assembly decided on April 29, 1791, to again permit soldiers to attend political clubs while off duty. Although this action had the effect of defusing some tension among the troops, its major impact was to widen the gap between officers and men. "Some of these clubs and a number of the more radical newspapers, were now calling for the wholesale dismissal of all noble officers on the grounds of their hostility to the new regime."[22] In addition, in some cases, the actions of clubs led soldiers to defy their officers openly. For example, the local political club at Saintes convinced the men of the 16th Infantry Regiment to disobey their officers' orders to return to their camp.[23] In essence, what the soldiers were demanding was direct democracy, something that is contrary to good order and discipline in any military.

Instead of just singling out the nobility, focus was increasingly being placed on officers per se. The fact that they had or did not have a noble background played less of a role. The mere fact that they were officers and tried to carry out their tasks with normal military discipline was enough to condemn them in the eyes of the more radical components of French society. In reality, the revolution never did make up its mind about the role and status of an officer.

> Were officers intended to exert authority and impose discipline when circumstances demanded it, or was such action to be construed as potential tyranny? Were officers expected to show individual initiative, or was their role to be restricted to the channeling and implementing of political instructions? Was the officer chosen for this expertise or for his political worth? Was he first and foremost a technician or a political appointee?[24]

In fact, when it came to political authorities, officers were often judged more harshly on their political shortcomings than was the case in other areas. To survive, an officer had to watch everything he said or did. The result was a further decrease in the number of officers. In April 1792 France went to war, by mid-1792, there were "regiments . . . almost without officers."[25]

One irony of the French Revolution was that in spite of its secular nature (and anti-Catholicism), chaplains continued to serve in the Revolutionary Army—at least, during this period.

> Many soldiers, setting out to face death, rejected de-christianization. They wanted to keep the priests who might help them with their pastoral ministrations at the supreme moment. These priests had often left home with them as chaplains in the volunteer battalions.[26]

Most priests were forced out, especially after Robespierre and the Jacobins came to power. In many instances, however, priests went underground and changed their roles. Some became ordinary soldiers, while others replaced officers or NCOs who were lost in battle. Later, the police were advised against permitting priests to join the military—even if they had a certificate of good citizenship and papers stating that they were no longer priests.[27] In time, their positions were used for other purposes—in the navy, for example, the position of "instructor" was created on warships in place of chaplain. The former had the task of teaching elementary subjects, such as reading, writing, and arithmetic, to the crew. The important point is that the continued push for the presence of chaplains in the armed forces underlined just how divided the army was. While many in the military welcomed the revolution and all that it stood for, others—including common soldiers—continued their ties to the old regime through their religious beliefs.

On April 20, 1792, France declared war on Austria. One of the few bright spots during this entire period was the outstanding way in which the army performed at the Battle of Valmy. At this small town, the French Army used massed artillery to stop the Prussian advance. As a consequence, "with the victory at Valmy the Convention declared that the *partie* was no longer in danger."[28] In spite of this victory, however, the military was still in a sorry state. The battle had been won with the assistance of thousands of volunteers, many of whom believed that once the battle was over, they were free to go home. Furthermore, while the importance of the victory at Valmy should not be downplayed, we must keep in mind that it was won by artillery, with the infantry playing a secondary role. How well the infantry would perform in a lengthy, pitched battle remained to be seen. In the meantime, something had to be done if the military was to be made into the effective tool the country's leaders in Paris hoped for.

On August 10, 1792, moderate Jacobins, also known as Girondists, took over power in the Assembly. These were men who favored a less radical approach to revolutionary politics. They wanted to preserve the king's position and they feared that the revolution could quickly get out of hand and lead to a number of excesses. They had seen the passion of the crowds that interrupted their debates in the Assembly, and they genuinely feared that if the masses were given too much power, it would be a disaster for France.

For the next ten months France would be locked in what amounted to a civil war between the Girondists and the more radical group of Jacobins. The latter placed primary emphasis on the masses and supported the *sans-culottes* (i.e., those from the lower classes) in their push for direct democracy.[29] By the first part of 1793 a civil war was under way, one in which those on the left, the Jacobins, were portrayed as part of a royalist conspiracy. The Girondists and the Jacobins locked horns in a no-holds-barred battle. On May 31, *sans-culotte* agitators surrounded the Convention and presented their demands, which included "the arrest of those Girondists most hostile to Paris, a tax on the rich, the creation of an army of revolutionary militants to punish suspects, the right of suffrage for *sans-culottes* only."[30] Two days later the matter came to a head. The convention was surrounded and the *sans-culottes,* backed up by the National Guard, demanded that the Girondists be handed over. The Convention meekly complied and, by acclamation, twenty-nine Girondist deputies were taken into custody and guillotined several days later. This marked the end of the moderate Girondist period and the beginning of the radical and terrorist Jacobin government. The proponents of direct democracy had won, a situation that would have the most serious implications for the army.

Creating Political Commissars

On October 1, 1789, Parliament created a military committee to provide deputies with expert advice. It soon became the focal point for military reform and was the source of many proposals for dealing with the army. Indeed, most of the laws relating to the armed forces that were promulgated in coming months originated in this committee. In addition, since the committee was staffed by individuals who had extensive military service, it was also useful as a way of establishing credibility in the eyes of the country's soldiers.

Although the presence of a special committee helped members of the National Assembly understand military problems, it did little to deal with issues of discipline and cohesion in the armed forces on a day-to-day basis. Recognizing these problems and the potential danger represented by the officers who had served the old regime, the Assembly came up with a unique two-pronged approach—at least, it was unique up to that point in time.

The country's leaders devised two structures to deal with political reliability questions. The first began in 1791, when the Legislative Assembly started sending agents to various army units to ensure the military's reliability. For example, in June commissioners, or representatives on mission—to use their formal term—were sent by the Assembly to administer the new loyalty oath to the National Assembly required of all soldiers.[31] When the Convention replaced the Assembly in September 1792, it reaffirmed the practice of sending out representatives on mission; in fact, it regularized and expanded the process by legislation passed on April 9 and April 30 in 1793.

In terms of their work, these representatives were primarily concerned about discipline and political reliability, at least in the beginning. By and large, they focused their attention on senior officers—many of whom were from the old royalist army. There was a constant concern that an officer would betray the revolution and work to restore the monarchy. Lynn noted the authorities' extreme sensitivity vis-à-vis officers' reliability when he observed:

> Even a hint of disloyalty propelled authorities into action against a suspected officer. A law of 24 June 1991 authorized generals to suspend any officer "whose conduct appears suspect." Later, the minister of war ordered his agents, the commissaries du conseil executif, to keep a close watch on individuals including "officers . . . who manifest opinions contrary to liberty, equality, and the unity and indivisibility of the Republic. In civic remarks could bring prison or death to the disloyal or unlucky. Association

with someone pronounced a traitor, particularly a traitor of high rank, could doom an officer to close scrutiny, if not to suspension or arrest."[32]

The authority of the representatives on mission was so great that they could appoint military officers (to replace the ones they had fired), they could requisition recruits from the local civilian population, and they could decree their own provisional laws. They also had the additional task of dealing with issues related to supply and morale, as well as to the unit's level of combat readiness. These representatives were members of the Assembly (and later of the Convention), and they were sent to various armies for short periods of time. Once they completed their inspection or carried out whatever tasks were assigned to them, they returned to their parliamentary duties in Paris. Some of them had served in the military previously, but by and large the majority lacked a military background.

The second category of political commissars came from the Ministry of War itself. In addition to its normal administrative duties, the Ministry of War played an important political role. It had responsibility for political education and also for eliminating officers who were considered politically unreliable. To carry out this function, the Ministry of War assigned officers—called commissars—to various units. As one instruction from the Minister of War dealing with the role of such officers noted, "The agent of the Council is the eye of the minister with the armies in order to discover all treasons, intrigues, and abuses."[33] These officers worked for the Minister of War and reported directly to him. They were especially important when it came to questions of political reliability. One such officer noted how a commissar reported:

> On officers he suspected, . . . since reporting on the political opinions of officers was part of his job. Spying may not be the word for it, but Calliez did report on the political acts and even the amours of Representatives to the Ministry and to the press.[34]

Military commissars also become directly involved in the military justice system. In May 1792, the Assembly created what was called a tribunal of military police, which was made up of three military commissars. They had a major impact on determining how nonjudicial punishment[35] was handled in the armed forces.

Military commissars also had responsibility for more mundane matters, which were crucial in keeping the army going. Commissars played a major role in areas like political education. As one military commissar was instructed by the Minister of War:

> One of the principal objects of your mission is the distribution of patriotic journals and the maintenance of our brothers in arms of the love of liberty which has made them win many victories, to warn them against the maneuvers of the aristocracy, and to unmask the false patriots who only want to win their confidence in order to betray the Republic.[36]

Because of the nature of their work, the activities of the representatives on mission and military officers serving as political commissars often overlapped.

The main difference between the two was that the military officers were assigned permanently to these units, while until April 1793 the representatives on mission came and went, depending on the whims of the Convention. After April 1973, representatives on mission were stationed at military units on a more permanent basis. Until that point, representatives might be there for several months or weeks, only to return to Paris—in essence, giving considerable power to the military commissars who enjoyed the power that came with always being present in a given unit.

The power of both sets of commissars was not limited to issues related to political reliability, however. They both bore responsibility for things like morale, motivation, and political socialization. Soldiers' living conditions, including food, uniforms, weapons, and training, figured high on their agenda. If problems arose, it was the task of the representatives on mission to resolve such issues and, if it was not possible to do so on the spot, then to refer the matter to the Convention for resolution. The same was true of War Ministry officers. Finally, representatives from the Assembly or Convention were often utilized to resolve high visibility problems. For example, when the Marquis de Lafayette had a change of heart and asked his troops to renew their oath to the monarch, the Assembly quickly dispatched several representatives to guarantee the army's loyalty.[37]

Commanders often resented interference by outsiders in what they saw as purely military affairs. Commissars were needed to help with morale, motivation, and political socialization, and this was fine. The problem was the representatives on mission. At least, the political commissars (who were not always loved either) were sent from the Ministry of War, were military officers, and understood something about fighting a war—unlike the dilettantes from the Assembly or Convention. However, many political commissars hated the military and saw a conspiracy behind every officer. Thus, from the standpoint of commanders, commissars were not necessarily easier to work with than representatives on mission. After all, the latter could sometimes be lulled into believing that everything was in order since they did not understand the details of life in the military.

Before the Jacobins came to power, neither the commissars nor the military representatives had enough authority to maintain the kind of order needed in the French military. Both commissars and military representatives helped make the military more reliable, worried about morale, and tried to motivate soldiers, but the revolutionary process destroyed military cohesion and order, as soldiers complained about officers abusing their authority, and ambitious officers criticized those above them as a way of moving up the promotion ladder. The situation was made even worse by the policy in some units of electing officers—a situation that placed tremendous power in the hands of subordinates. In order to be elected, an individual had to curry favor with his fellow soldiers. As a result, reliance on meritocracy as a way of rising through the ranks became less important.

Because of the overlapping nature of their authority, representatives on mission sometimes resented the military commissars in the field and occasionally found themselves at logger-heads with them. The War Ministry representative might decide that an individual was perfectly suited for a given job, while the representative on mission might oppose the individual's appointment to that position. There were occasions when the two men worked together and complemented each other, but when a difference of opinion arose, it was the representative on mission who won out since he carried the authority of the National Assembly and later the Convention. These structures represented the highest executive (and legislative) authority in the land. In the process, however, the activities of the military commissars often irritated the representatives from the country's legislature. As Bertaud noted, "The representatives on mission disliked these competitors and made their feelings known."[38] In addition to the obvious jurisdictional problems, military commissars were also resented because many from the Assembly and later the Convention were generally suspicious of all French officers. Having military officers serving as political commissars—even if they were sympathetic to the revolution and everything that it stood for—was more than many radical revolutionaries could stomach.

The Army and the Jacobins

The attitude of Maximilian Robespierre, the leader of the Jacobins, and his colleagues toward the army was contradictory. On the one hand, they had a paranoid fear that the army, and particularly its officer corps, would attempt to turn the clock back and reinstate the previous regime. Indeed, many of them saw the officers (especially the more senior ones) as closet monarchists, who would be only too happy to see the House of Bourbon back in power. On the other hand, the Jacobins realized that they needed the army. Without it, France's enemies would soon be in Paris.

Few events in the French Revolution had a greater impact on the country in general and the military in particular than the April 4, 1793, decision by General Charles Dumouriez to defect to the Austrians together with the Duc de Chartes, who was to become the future Louis-Philippe. Dumouriez, a likable soldier, had held a number of high-level posts in the Revolutionary government, and as one of the main architects of the victory at Valmy he had become a national hero before taking command of French troops in the North. After suffering a couple of defeats at the hands of the Austrians, he turned on the revolution. He not only surrendered the Low Countries to the Austrians, he announced his intention of marching on Paris and restoring the monarchy. When the Convention sent representatives, together with the country's defense minister, to confront him, he had them arrested and subsequently turned them over to the Austrians. In the end, his troops refused to go along and he defected to the Austrians. Paris was ablaze with indignant cries of betrayal—one of the old generals, a hero of the revolution, had betrayed it!

The defection of Dumouriez in 1793 became one of the political cause celebres of the revolutionary decade because he not only surrendered a number of border forts to the Austrians, but also tried to turn his army against Paris.[39]

What did this say about the other generals, not to mention other senior officers? How could one guarantee that this might not happen again?

Continued concern over political reliability, combined with ever present evidence that insubordination was becoming more common and the fact that the Jacobins were even more radical than others in their hatred toward and suspicion of the actions of officers, meant that the military would get a new and more careful review. One thing was apparent to the country's leaders—France had to have a strong military since it was surrounded by enemies, the vast majority of whom would be only too happy to restore the old monarchy. But how to do it? Clearly, the old approach had not worked. Political commissars, whether of the military or civilian type, were not sufficient. If they were to be employed in the future, major reforms were in store.

Given their fear of a military take-over, as well as their resentment at what many of the Jacobins saw as military interference in political matters, one of the first things the Jacobins did was to revamp the War Ministry. They wanted to be certain that the revolution would command the sword and not the other way around.

On October 10, the Convention proclaimed the existence of a provisional government that would be "revolutionary until the peace."[40] This new revolutionary government would be different from governments of the past. The Committee on Public Safety was given expanded power and took the lead in reestablishing central control over all parts of French society—including the army.

The men of 1793 enlarged the scope of the royal "extraordinary" to turn public safety into a regime which suspended constitutional laws and was entirely directed toward rebuilding a strong central government which would be obeyed unquestionably. Public need was placed above the law, and the state's arbitrariness accepted in the name of efficiency.[41]

In essence, France had now entered the period of terror—a time when no one was safe. In addition to regaining central control over the country, the provisional government also had the advantage in that it served as a means for satisfying public dissatisfaction. As Maurois pointed out,

But the Government, without means to assure a supply of food, provoked resentment. "The time has come," said Madame Roland, "which was foretold, when the people would ask for bread and they would be given corpses." Then it was that informing became an act of civic duty, the guillotine an altar of virtue. For fourteen months the Revolutionary Tribunal sat without recess.[42]

The country's armed forces were not spared this reign of terror—generals and admirals, as well as common soldiers, found themselves facing the guillotine for nothing more than "thinking" counterrevolutionary thoughts. Indeed, now the search was not just for the guilty but for those who "might" be guilty. Even doing nothing for the revolution could be considered grounds for punishment. As one writer put it, "Between the spring of 1793 and fall of Robespierre in July 1794, political loyalty and reliability outweighed all other criteria for military command."[43] France had entered the darkest days of its revolution.

One of the most immediate effects of the advent of terror was that it intensified the problem of inexperience on the part of career military personnel. To take only one example, during the period from 1792 to 1799 the army had a total of 1,378 generals. According to French military records, during that time 994 of them were brought before the tribunal of justice, with many of them being executed.[44] Needless to say, this led to a tremendous turnover at the top. As a consequence, not only was the kind of experience an army needs lacking among senior officers, many individuals actually refused promotions, fearing that they, too, would quickly fall victim to the revolution and its excesses.

To further tighten control over the army, on December 4 the government issued a directive according to which the Minister of War and all generals, admirals, and army corps were placed under the authority of the Committee on Public Safety. "All constituted bodies and public functionaries came under its direct supervision."[45] Military autonomy was limited to the point where all administrative memorandum now had to pass through the Committee's hands. Shortly thereafter, the Ministry itself was dissolved. The Committee on Public Safety had become the country's "super" political commissar.

To make it clear who was in charge, one of the first things the Committee did was to dismantle the Ministry of War. The War Ministry's functions were divided between several of the country's twelve executive commissions. The fourth commission (trade and consumer goods) was responsible for feeding and outfitting the army; the fifth for public works; the seventh for transport, post, and repair; the ninth for the organization and movement of the ground armies; and the eleventh for weapons, powder, fortifications, and weapons production.[46] The coordinating function remained in the hands of the Committee on Public Safety, which had a number of military specialists on its staff. Centralized control over military matters by professional military officers was a thing of the past. For a variety of reasons, the new system seemed to be working—as France won major victories in the field in May and June of 1794.

The power of military commissars was now greatly limited. They remained in place, but when it came to tasks like ensuring political reliability, raising morale, motivating soldiers, or making certain that political indoctrination work was proceeding apace, commissars were clearly subordinate to civilian authorities. They were not considered sufficiently reliable by Jacobin authorities.

The situation with regard to representatives on mission was quite different. First, almost without exception these individuals were committed Jacobins. They took their jobs very seriously and believed it was up to them to instill a deep sense of patriotism and commitment to the revolution on the part of the average soldier—and they worked hard in carrying out their tasks. They had a number of tools available to them for this purpose.

In practice, the Convention devoted considerable expense to political socialization by ensuring that the soldier in the field knew what was going on in Paris from newspapers and journals that were made available to them free of charge. In addition, festivals that focused on a patriotic theme were organized, and political slogans were commonly used throughout the army. Even theater was employed to inspire the soldiers and to ensure that they accepted the messages of the new regime.

Of all the techniques the Jacobins and their predecessors utilized, one of the most effective was songs. "Songs were certainly more popular with the troops and were probably more important in promoting patriotism and revolutionary values."[47] The important point was that under the Jacobins, it was almost impossible to escape the politicizing influence of the representatives on mission. Everything they did was full of the revolution, its goals, and its need for sacrifice. And this process seems to have been effective.

> From the soldiers' own diaries and letters it is clear that this propaganda bore fruit, that the soldier of the Year II, more than his counterpart under the Directory in the line army of 1789, did think politically and had at least some basic inkling of the nature of the cause in whose name he was enlisted.[48]

One thing that the Jacobins quickly realized was that despite the progress they were making in politicizing the troops, they were also faced with a disintegrating military.

The situation was made worse by the constant involvement of the Jacobin Clubs around the country in military affairs. "Both in the sections of the larger towns the local Jacobin clubs throughout France took it upon themselves to provide political surveillance for the troops reporting on their conduct, their public spirit, and on the reputation of their officers."[49] In short, not only did the army have to face increased political involvement in its internal affairs by senior representatives from Paris, it was also forced to deal with greater interference by public organizations, most of which knew very little about the military—except for having the belief that it was undemocratic and that its officers were probably enemies of the revolution.

There was a real danger that the ideas of "liberty, equality, and fraternity" that had been so widely spread around the country would destroy completely the idea of a hierarchical military. If everyone was equal, then why should soldiers obey the orders of their officers—especially if those orders could get them killed or maimed? Paris ran the risk of having the most reliable, most dedicated, and most useless military anywhere in Europe.

Before we discuss how the Jacobins dealt with the problem of military disintegration, it is important to understand just how serious it was. Take, for example, the surrender of the key naval port of Toulon on August 27, 1773, to the British—not long after Robespierre and the Jacobins had come to power.[50] In August, Toulon rose in revolt against the Convention, declared its allegiance to Louis XVII, and opened its port to the British fleet. Not only did the French Navy lose seventeen battleships in the process, the fact that the city had surrendered to the British without having fired a single shot in its own defense was something that Paris found hard to take. Cormack argues that "News of the disaster led directly to the formal inauguration of the Terror in Paris."[51]

Although there are different versions as to what actually happened, the French Navy itself was divided, with some favoring the Jacobins and others loyal to the monarchy. For average Frenchmen or women, what stood out most clearly was their increasing dissatisfaction over the extremism of the Jacobins, especially the hundreds or thousands of victims the guillotine claimed daily. As a consequence, when the question came up of whether or not to open the city to the British, navy personnel did not know whom to obey. Senior naval officers had good grounds for disliking the Jacobins. Many of their colleagues had been imprisoned because of denunciations instigated by the Jacobins. Had there been clear direction from Paris or had the city not been so strongly opposed to the Jacobins, the military probably would not have given up the city without a fight. As Cormack put it, "Naval commanders in Toulon probably welcomed the collapse of Jacobin power, but they did not contribute to it."[52] On the other hand, it is important to keep in mind that just as military action on one side or another can tip the balance in a political conflict, not acting can have the same effect. The failure on the part of the French Navy to resist local efforts to desert revolutionary France made surrender of the port a foregone conclusion. Although it was eventually recaptured on December 19, the ships had been lost and France's honor deeply wounded.

The solution selected by the Jacobins for dealing with this spreading disorder and chaos was most unusual. The representatives on mission, who up to then had devoted a considerable amount of time and effort (especially those who were Jacobins) toward undermining the authority of officers and other authority figures, became the primary vehicles for enforcing discipline—when necessary, as it usually was—in a thoroughly ruthless fashion. They worked directly under the authority of the Committee on Public Safety. Indeed, compared with the prerevolutionary period, the representatives on mission were far more draconian in enforcing discipline and order than most royalist officers ever were. The latter could be brutal, but not to the degree that the representatives on mission were under the Jacobins.

It was in Brest that the role played by the representatives on mission in restoring order was most evident. On September 13, shortly after the news of the surrender of the fleet in Toulon reached Paris, a mutiny took place in Brest. Even the dispatch of a representative on mission did little to bring order and convince the sailors to go back to their ships.

Faced with mutiny and fearful that the situation could become another Toulon, the Committee of Public Safety sent out one of its "super" political commissars—and a naval expert—Jeanbon Saint-André, to deal with the crisis. From the very beginning it was clear that Saint-André was very serious and prepared to use whatever measures it took to restore order. As he put it in his report to the Committee, "It was necessary to destroy, to annihilate at any price, to deliver to our most cruel enemies this bulwark of our security."⁵³

Jeanbon arrived in Brest on October 7 and spent the first couple of days inspecting the fleet. A few days later, the admiral in charge, Vice-Admiral Morard de Galles, and several of his deputies were relieved of command. Jeanbon also stripped several captains of command and ordered the arrest of others. A number of the captains and men of other ranks were sent to Paris, found guilty of misconduct, and guillotined. Other sailors, who expected to be praised for standing up to authority, soon found themselves in irons—and some were likewise guillotined.

> Several gunners, sailors and marines have also been placed under arrest. This severity was just, it was necessary; because discipline must reign, . . . this was not recognized and the obedience of subordinates to their commanders, which is only the obedience to the nation itself which named them, was trampled on.⁵⁴

From Jeanbon's standpoint—as well as that of his fellow representatives on mission in Brest—the actions of mutineers were not a sign of loyalty to the revolution but part of a sinister plot aimed at destroying it. To satisfy pressures coming from the left, most of the few aristocrats still in the fleet also were dismissed.

Jeanbon addressed the fleet and argued that while all of the aristocratic officers—who were the source of the mutiny—had been removed, "from ship's boy to admiral, order must reign."⁵⁵ Further efforts at mutiny would not be tolerated.

> Do not doubt that the blade of the law will strike all conspirators without pity. The Nation wants only faithful servants: it will punish insubordination and cowardice with firmness; as great in its rewards, it will be more severe and inflexible in its punishments.

To back up such rhetoric, Jeanbon introduced a new penal code, one that contained harsh punishments for violations of naval regulations.

> Sailors not obeying orders quickly would suffer four days in irons. Refusal to obey orders, if accompanied by threats, would result in flogging or five years imprisonment. Raising a hand to a superior warranted the lash, while more serious assault was punishable with death. Death was also the penalty for inciting mutiny.⁵⁶

Sailors no longer had the right to question orders—the idea of petitions (so popular just a year earlier) was dismissed out of hand. Jeanbon delivered the same message to officers—many of whom had hoped to hop over those above them by denouncing them for political crimes. This was now the "sailors' navy" and, as such, it would not tolerate sailors who could not or would not obey orders. The only explanation for such actions in Jeanbon's mind was betrayal of the revolution.

It was not long before Jeanbon and his colleagues had expanded their terror campaign to the city of Brest itself—including all of its citizens. This meant that some members of the local Jacobin society also found themselves in trouble. Jeanbon then returned to Paris to report to the Convention on his actions.

Soon, however, Jeanbon found himself going back to Brest, having been charged by the Convention with the task of building up a modern, combat-capable fleet—one that could stand up against the British. Jeanbon quickly divided his task into four areas; the construction, fitting out, and repair of warships; finding sufficient manpower; making the arsenal operative; and training crews. Jeanbon and the representatives on mission went to work—requisitioning materials and supplies from civilians when necessary—but in the meantime they devoted considerable energy and effort to achieving their goals. Regarding the civilian workers, Jeanbon imposed the same kind of discipline on them as he had on the fleet's sailors. Precise starting hours with roll-calls were introduced. Workers were also placed under surveillance and those who did not carry out their jobs were called to task for their actions. If they did not correct their behavior, they could find themselves in jail, if not worse.

When it came to selecting officers to lead the fleet, Jeanbon took a page from Cromwell's diary by placing talent ahead of ideological purity. This was clearly the case with the Count Villaret de Joyeuse. He was of noble birth, and although he had the support of the sans-culottes, the fact that such a person would be promoted to the position of commander in chief of the fleet was unusual in those days of ideological fervor. As far as Jeanbon was concerned, officers who "played the game" and proved that they were good sailors could expect to be rewarded—as long as they did not rock the ideological boat. On the other hand, those who turned out to be poor sailors—regardless of their ideological orientation—would find themselves fired, and in some cases they ran the risk of being forced to account for their actions before a court martial. Ideology, yes, but ideology without technical competence had no place in Jeanbon's navy.

Terror in France—and that includes the navy—came to a sudden halt on July 27, 1794, the day of Thermidor, when Robespierre's Jacobin government came tumbling down. Terror was an effective form of government as long as those running the government believed it was serving a useful function (getting rid of those who threatened its existence). Over time, however, the French became increasingly concerned about the excesses of the revolution—some 14,000 people had been executed![57] Resentment was growing. The law of June 10, 1794, that Robespierre pushed through—a law that deprived deputies to the convention of their immunity from arrest—was a clear warning. They could be the revolution's next victim. Robespierre had gone too far. He was arrested and, like his many victims, guillotined!

Thermidor and the Military

Thermidor had a major impact on the French military. To begin with, the Thermidorians stopped the terror—after getting rid of Robespierre and his closest colleagues. The Ministry of War was reestablished in place of the commissions the Jacobins had set up and was given back authority over the armed forces.

The importance of politicization in the military decreased significantly. The number of political tracts sent out from Paris to the troops quickly diminished. Those that did appear had a much more patriotic—rather than ideological—tone to them. Similarly, troops were more isolated from civilian society than had been the case at any time since 1789. The military press also became less ideological—more like those available in present-day militaries—as the revolutionary rhetoric disappeared.

On a personal level, an officer's political opinions gradually became his own private affair—as long as he did not challenge the established order. Military courts, which were now made up exclusively of military officers (instead of the mix of civilian and military common to the revolutionary period), began to focus less on political crimes—which had been their main preoccupation under Robespierre—and to worry primarily about the offenses that concern militaries all around the world. At the same time, hundreds of officers who had been imprisoned under the revolutionary government were rehabilitated and returned to service.

Another change that was directly related to the effort to introduce order into the military was the law of April 3, 1785, which significantly modified the policy of electing officers. Another law—that of November 10, 1795—put the whole issue in the hands of the Directory, which had become the country's senior executive authority in 1795. Electing officers might be one of the highest forms of direct democracy. However, when it came to running a military, electing officers was an anachronism and continually undermined military order and discipline.

Representatives on mission were still sent out from Paris to check on military units—but their powers were greatly reduced. In fact, they were no longer referred to as representatives on mission but as military commissars. This put them on the same level as the military officers who were still worrying about things like morale, motivation, and political education.

> Their unlimited powers were removed, their financial resources controlled, and their right to overrule generals and to appoint military officers withdrawn. They were there to encourage the zeal of the officers rather than to give them orders; the political arm was no longer supreme.[58]

One interesting footnote to history: Alan Schom, in his biography of Napoleon, claims that the leader's brother Lucien, who was constantly getting into trouble, was appointed a political commissar in late 1795.[59]

One thing was clear—regardless of whether they were civilians or military officers, no longer would political commissars have the kind of authority that they enjoyed in the past. For purposes of this study, they were assuming a role more like that of a political officer—an individual whose goal was to motivate soldiers, to worry about their morale and personal concerns, to socialize them, and to assist the commander in the maintenance of a high level of combat readiness.

Conclusion

In ten years, France had gone through major changes. The army that took the field in 1775 bore little resemblance to the one of 1789. Formerly monopolized by the nobility, the officer corps was now open to all segments of French society. Furthermore, the French soldier was no longer the robot that he had been under the old regime—he had rights (even if less than he thought he had in 1793), and the government was expected to respect them (even if Napoleon would forget that fact on more than one occasion).

In terms of political commissars, France had also undergone major changes. Originally introduced as a means of ensuring the political reliability of an officer corps that appeared closely tied to the old regime, political commissars over time became much more than that. Indeed, as the example from Brest shows (and many others could be supplied), such individuals ran the military. The fact that the French fleet was repaired and put into good order in such a short time was acknowledged by almost everyone—including senior naval officers—to have been a result of the work of Saint-André. Regardless of whether these men were functioning as "super" political commissars, as representatives on mission, or as military commissars, they played an invaluable role. Without their hard work, a lot of French soldiers would have gone without clothes, food, or other supplies, and the level of morale would have been considerably lower.

This is not to suggest that the French managed to overcome the contradiction between political reliability and military expertise. In fact, the latter suffered considerably, even if the French army was able to win some important victories by its deployment of troops on a massive scale. The bottom line, however, was that when push came to shove, the Assembly and later the Convention considered political reliability to be of primary importance. Without that, there would not have been a revolution—the country would have quickly reverted back to being a monarchy.

What is of equal importance is that once the major threat was over—once those in power decided that the country's army and navy were sufficiently loyal and reliable—political commissars were quickly withdrawn. They did not totally disappear, but their functions soon began to closely resemble those of a political officer.

Notes

1. Throughout this chapter, officers will be referred to as "he" because very few women served in the French military at this time.

2. As cited in John A. Lynn, *The Bayonets of the Republic* (Boulder, Colo.: Westview, 1996), 44. Alan Forrest gives slightly different figures: 113,000 infantry, 32,000 cavalry, and 10,000 artillery. Alan Forrest, *Soldiers of the French Revolution* (Durham, N.C.: Duke University Press, 1990), 27.

3. Jean Paul Bertaud, *The Army of the French Revolution* (Princeton: Princeton University Press, 1988), 17.

4. Bertaud, *The Army of the French Revolution*, 19.

5. Forrest, *Soldiers of the French Revolution*, 27.

6. Bertaud, *The Army of the French Revolution*, 19, 21.

7. S. F. Scott, "The French Revolution and the Professionalization of the French Officer Corps, 1789–1793," in Morris Janowitz, *On Military Ideology*, ed. Jacques van Doorn (Rotterdam: Rotterdam University Press, 1971), 8.

8. Bertaud, *The Army of the French Revolution*, 23.

9. Bertaud, *The Army of the French Revolution*, 25.

10. Forrest, *Soldiers of the French Revolution*, 17.

11. John Ellis, *Armies in Revolution* (London: Croom Helm, 1973), 80.

12. Forrest, *Soldiers of the French Revolution*, 20.

13. Forrest, *Soldiers of the French Revolution*, 18.

14. The best explanation of this event in English is in Forrest, *Soldiers of the French Revolution*, 22.

15. Francois Furet, *Revolutionary France 1770–1880* (Cambridge: Blackwell, 1988), 122.

16. Scott, "The French Revolution and the Professionalization of the French Officer Corps," 24.

17. William S. Cormack, *Revolution and Political Conflict in the French Navy, 1789–1794* (Cambridge: Cambridge University Press, 1995), 141.

18. Bertaud, *The Army of the French Revolution*, 34.

19. Lynn, *The Bayonets of the Republic*, 121.

20. Forrest, *Soldiers of the French Revolution*, 20.

21. Cormack, *Revolution and Political Conflict in the French Navy*, 141.

22. Scott, "The French Revolution and the Professionalization of the French Officer Corps," 23.

23. Scott, "The French Revolution and the Professionalization of the French Officer Corps," 25.

24. Forrest, *Soldiers of the French Revolution*, 54.

25. Scott, "The French Revolution and the Professionalization of the French Officer Corps," 25.

26. Bertaud, *The Army of the French Revolution*, 185.

27. Bertaud, *The Army of the French Revolution*, 186.

28. Lynn, *The Bayonets of the Republic*, 111.

29. *Sans-culottes* refers to that segment of the French population that did not wear fine trousers. In reality, this meant the poor and disenfranchised.

30. Furet, *Revolutionary France, 1770–1780*, 127.

31. Representatives on mission are sometimes described as deputies on mission or commissioners, depending on the author involved.

32. Lynn, *The Bayonets of the Republic*, 83.

33. As cited in Lynn, *The Bayonets of the Republic*, 84.

34. Lynn, *The Bayonets of the Republic*, 85.

35. *Nonjudicial punishment* refers to disciplinary actions taken short of a formal court martial proceeding. For example, minor violations of military discipline such as excessive drinking or being late for formations are normally handled by nonjudicial punishment.
36. As cited in Lynn, *The Bayonets of the Republic*, 126.
37. See Lynn, *The Bayonets of the Republic*, 78.
38. Bertaud, *The Army of the French Revolution*, 194.
39. Forrest, *Soldiers of the French Revolution*, 24.
40. As cited in Bertaud, *The Army of the French Revolution*, 157.
41. Furet, *Revolutionary France, 1770–1880*, 129.
42. Maurois, *A History of France*, 314.
43. Scott, "The French Revolution and the Professionalization of the French Officer Corps," 26.
44. Lynn, *The Bayonets of the Republic*, 81.
45. Bertaud, *The Army of the French Revolution*, 157.
46. See Andreas Kieselbach, "Entwicklung des französischen Kriegsdepartements von 1785 bis 1814," *Militärgeschichte* 1 (1989): 24–27.
47. Lynn, *The Bayonets of the Republic*, 142.
48. Forrest, *Soldiers of the French Revolution*, 116.
49. Forrest, *Soldiers of the French Revolution*, 90.
50. The primary source for this incident and Paris's response is Carmack, *Revolution and Political Conflict in the French Navy, 1789–1794*, 173–214.
51. Cormack, *Revolution and Political Conflict in the French Navy, 1789–1794*, 213.
52. Cormack, *Revolution and Political Conflict in the French Navy, 1789–1794*, 187.
53. As quoted in Cormack, *Revolution and Political Conflict in the French Navy, 1789–1794*, 215.
54. As quoted in Cormack, *Revolution and Political Conflict in the French Navy, 1789–1794*, 252.
55. Cormack, *Revolution and Political Conflict in the French Navy, 1789–1794*, 252.
56. Cormack, *Revolution and Political Conflict in the French Navy, 1789–1794*, 252.
57. Maurois, *A History of France*, 324.
58. Forrest, *Soldiers of the French Revolution*, 121.
59. Alan Schom, *Napoleon Bonaparte* (New York: Harper, 1997), 233. As far as I can determine, either Schom misunderstood the concept of a political commissar or he is wrong. Reading several autobiographies of soldiers who served under Napoleon, I was struck by the failure of these men to even mention the institution—and at least one of these accounts is very detailed. See *The Memoirs of Baron De Marbot*, Arthur John Butler, trans., in two volumes (London: Longman's Green, 1892); *With Napoleon in Russia, 1812* (London: Folio Society, 1969); and *Life in Napoleon's Army: The Memoirs of Captain Elzear Blaze* (London: Greenhill, 1995). It thus appears that the concept of a political commissar, as it was understood during the French Revolution, did not survive into Napoleonic times.

4

Commissars in the Red Army

Only irreproachable revolutionaries, staunch champions of the proletariat and the village poor, should be appointed to the posts of military commissars, to whom is handed the fate of the Army.

—Fifth Congress of Soviets (1918)

THE TAKEOVER OF THE czarist government by Lenin and his Bolshevik colleagues led to a civil war. The situation was very serious. Prior to the seizure of power, the Bolshevik's main purpose had been to destroy the Imperial Army by undermining discipline and creating a chasm between officers and enlisted personnel. Only if the army collapsed could the Bolsheviks take power. After years of war and Bolshevik agitation, this is exactly what happened. The Imperial Army disintegrated as hundreds of thousands of troops fighting at the front toward the end of World War I voted with their feet—by throwing down their rifles and returning to their homes and villages. The seriousness of this situation was highlighted by commissars, who had been appointed by the provisional government. These czarist commissars noted that the soldiers, "armed and in good health and high spirits, are certain they will not be punished."[1] The old army was in chaos. Its complete collapse—and with it, the provisional government—was only a matter of time.

But now a new danger loomed on the horizon, one that threatened to destroy the newly won gains of the communist revolution. Members from the ancien régime had formed themselves into an army—the so-called Whites—whose goal was nothing less than the overthrow of the Bolshevik regime and a restoration of the monarchy. This was no idle threat. The Whites had many of the country's best and most competent military officers in their ranks and if something wasn't done quickly, everything that Lenin and his comrades had worked for would come crashing down around their heads. Bolshevism would be a short-lived experiment at best. What to do?

The Bolsheviks needed officers from the old Imperial Army. They did not have anywhere near enough trained experts to staff their new Red Army. Even with the limited technology of the day, it took several years to train officers capable of leading the kind of armed forces the Bolsheviks required. The simple fact was that the only source of trained, competent officers was those who had served in the Imperial Army. Some of these officers had joined the Bolsheviks voluntarily. But there were not nearly enough of them to staff the army. Additional sources of commanders and other line officers would have to be found.

Among his other talents, Lenin was a great pragmatist. He, better than many in Russia, realized that while most officers from the former Czarist Army did not support the values of the new regime, they were indispensable to its survival. Without them, all of the ideological dedication in the world would be useless. Unfortunately, in many cases, their value system was 180 degrees opposite that of the communists. Many of them remained committed to the monarchy and the Russian Orthodox Church.

As a result, the Bolsheviks took a page from the French Revolution—an event that was constantly on their minds and served as a great inspiration to them. They introduced political commissars—individuals who would have a dual task. First, these individuals would be responsible for ensuring that potentially disloyal line officers were reliable. If an officer acted in a manner at odds with the goals or policy of the party, it was up to the commissar to either inform his superiors or, if necessary, dispatch the guilty party himself. A second task for the political commissar was political socialization. He had to make certain that those who were inducted into the Red Army, as well as those who would subsequently join it, accepted the new Bolshevik value system—the only way their loyalty could be ensured over the long run. Officers and NCOs who willingly supported the new regime would make for a much more efficient military and more stable regime. Watching every action of line officers not only got in the way of military matters, it also created problems of uncertainty with regard to the military, thereby making the construction of a viable system of civil-military relations very difficult. Finally, political commissars would be responsible for the more mundane affairs of morale and motivation.

Making Use of Officers from the Czarist Army

One of the ironies of history was the contradictory position the Bolsheviks found themselves in after seizing power. After all, their previous goal had been to destroy the army. For example, on November 23, 1917, the party leadership had issued a decree on the army's gradual demobilization. Several weeks later the

Bolsheviks issued two new decrees, "On Elective Command and Organization of Discipline in the Army," and "On the Equalization of Rights among Soldiers." The first decree outlawed the old hierarchical command system. The second ordered the end of the use of all insignia, decorations, and officer organizations. In addition, the election of officers was introduced.[2] "By the decree on command, the regimental, battery, and squadron commanders were to be elected by the existing committees; higher commanders were conferred by the nearest higher committee."[3] Needless to say, the idea of electing officers—one that was taken from the French Revolution—created chaos. Just as in the French case, the belief that officers served at the pleasure of their troops was a recipe for disaster. In essence, the Bolsheviks achieved their goal—the end of the Imperial Army as a cohesive, competent fighting force.

The chaos that was the Imperial Army was evident in the overwhelming defeat that it suffered at the hands of the Red Guards and the Germans in separate conflicts in early 1918. Clearly, something would have to be done to build up a meaningful military. The pressure for reconstructing the military became even more intense when it appeared that Japanese troops were ready to pounce on Siberia, while the Whites became active in a variety of places around the country. The Soviet state stood in peril and it was obvious to Lenin that without a good army, the revolution would soon be nothing but a distant memory.

Before proceeding further, a definition is in order. Insofar as officers from the Imperial Army were concerned, those who had the technical expertise that the Bolsheviks needed were referred to by the communists as "military specialists." Utilizing these individuals was not only necessary, it was "a natural and inevitable means for creating detachments to repulse the attacks of the White Guards," as Sergei Gusev, a senior communist official, put it.[4]

Military specialists were employed in a number of areas. One of the first was as advisers. Using military specialists in this capacity enabled the Kremlin to draw on their knowledge, while at the same time, the Bolsheviks avoided the need to use them in operational areas, where they would come into direct contact with troops and where the danger would be greater that they could undermine the Bolshevik military effort. To get some idea of just how widely utilized such officers were, Mikhail Bonch-Burevich, a former general in the Imperial Army, was the Bolsheviks' primary military adviser in 1918. In addition, "practically all, of the twenty-six members of the Higher Military Council in 1918 had served in the Tsarist [Czarist] Army—13 of them had been members of the General Staff."[5]

Another area where military specialists could be utilized—and isolated from operational matters—was as teachers or professors. Thousands of such individuals, including former czarist officers, shared the commitment to a communist future and were willing to serve the Bolsheviks, but they lacked the necessary technical or military skills. They were in serious need of technical training. These were the kind of officers the Bolsheviks wanted most—individuals on whom they could rely. The lack of skilled officers led to the issuance of a decree entitled "The

Basic Conditions for Accelerated Courses for the Preparation of the Command Staff of the Workers and Peasants Red Army," on February 10, 1918. These courses were aimed at "preparing instructors for the infantry, cavalry, artillery, engineering forces and machine gun units." The length of the courses was set at either three or four months, depending on the subject area. Needless to say, in order to be accepted as a student, the individual had to convince Bolshevik authorities that he was politically reliable.

Despite the presence of these individuals, however, no alternative existed to reliance on military specialists. As a Soviet writer observed, ". . . for the general military schools for the workers, a tremendous number of instructors were required. It would have been difficult to solve that task without the old military specialists."[6]

In addition to short-term courses, military specialists also worked as instructors for long-term courses. Military specialists were given the task of reorganizing the entire educational system, to cite only one example. The General Staff Academy was placed under the command of A. K. Klimovich, who had been an officer in the Imperial Army. The new school opened its doors on December 8, 1918. In the first course, which lasted seven months, 183 students were enrolled. Other academies were also created and dealt with topics such as artillery, engineering, medicine, and naval affairs. Military specialists were indispensable: "Instructors at the Academy were drawn from former professors and teachers at the old Nikolaevskiy Academy of the General Staff and commanders of the Red Army, who had combat experience and as a rule, higher military education."[7] There were few areas where military specialists were more dominant than in the educational sphere. According to one source, "90 percent of the instructors and line officers serving at military academies, higher schools, accelerated and short-term courses were military specialists."[8]

The list of senior officers teaching long-term courses, in particular, reads like a Who's Who of Russian military thinkers. Among those who taught at the General Staff Academy were N. A. Danilov, V. F. Novitsky, N. A. Suleyman, and A. A. Svechin. The latter had an especially profound impact on early Soviet military thinking.

By the time Trotsky was appointed commissar of war on March 4, 1918, it was clear that a major reorganization was necessary if the Bolsheviks hoped to field the kind of army necessary to defeat the Whites. To Trotsky, the issue was simple—the role to be played by military specialists had to be expanded. This meant using them not only as staff officers, advisers, instructors, and educators but in operational, line positions as well. He argued that such individuals would play a limited role as line officers and that whatever the ideological wishes of the Bolsheviks might be (i.e., to utilize only those who were fervent believers in the communist ideology), the new regime had no choice but to use them.

Yes, we are utilizing military specialists because the task of Soviet democracy does not dispose of the technical forces we need for our historical work. . . . Military specialists will supervise technical matters, purely military questions, operational questions, combat issues.... At the current point in time, we have no alternative. It is important to remember that besides enthusiasm ... technical knowledge is also necessary.[9]

Trotsky's effort to enhance the roles played by military specialists ran into stiff opposition. Using them as advisers and educators was one thing, but giving them operational command was more than many Bolsheviks could stomach. In this context, it is important to keep in mind that during the period when the Bolsheviks had been agitating within the Imperial Army, these very same individuals—officers in the old army—had been singled out for the most vile criticism and attacks. They were the heart of all that was evil about the old system, and now Trotsky was suggesting that the new Red Army not only needed them but was dependent upon their service.

Two of the most outspoken opponents of the use of military specialists were Nikolay Krylenko and Nikolay Podvoysky. These two committed communists maintained that military specialists "cannot understand any other kind of war than a war utilizing huge masses of regular troops ... a situation that always demands bureaucratic centralization." They were worried that if military specialists were appointed, they would ignore Marxist ideology and pay little attention to the Bolshevik idea of doctrine: one that placed a premium on maneuver and the offense. Instead, Krylenko and Podvoysky argued that these officers would cling to the old ways.[10]

Not surprisingly, political reliability remained a matter of prime concern—and with reason. There were occasions when "military specialists joined the Red Army with the purpose of undercutting its combat capability from within." The actions of these "traitors caused considerable losses."[11] Regardless of the concerns raised by Krylenko, Podvoysky, and others, in the end Trotsky won. The Fifth All-Russian Congress of Soviets adopted a special resolution stating that "in order to create a centralized, well educated and supplied army, it is necessary to utilize the experience and knowledge of the many military specialists from the ranks of officers of the former army."[12]

To get an idea of just how serious the situation was, consider the following. The Soviet government had plans for the creation of some 60 divisions. However, by the end of 1918 the Red Army had only 1,774 Red Commanders (communists who had gone through the short-term training courses and were now serving as commanders). Those in charge of the military argued that 55,000 military specialists were required "just to fill command positions." By the end of 1918, 8,000 of these individuals had volunteered, while another 5,000 had been drafted.[13] And this was just the beginning. By the end of the civil war, Soviet sources state that, of the 130,000 officers in the Red Army, between 70,000 and 75,000 were military specialists.[14]

During the civil war, military specialists played key roles not only in educational institutions, but as staff and operational officers as well. When it came to operational commands, they dominated the senior level positions. For example, of the twenty officers who commanded the Eastern or Southern Fronts between 1918 and 1920, seventeen, or 85 percent, were military specialists. The same was true of other positions. All of the chiefs of staffs of fronts, 82 percent of the army commanders, and 83 percent of the chiefs of staff of armies had served in the Imperial Army.[15]

Controlling the Military

In order to ensure the reliability of these officers, the Bolsheviks took two important steps. To begin with, all military specialists who joined the Red Army (whether voluntarily or involuntarily) had to go before a special "Attestation Commission." It was the task of this commission to evaluate each officer's political reliability. The added danger was that if an officer were judged to be unreliable, he might be shot out of hand for fear that he would join the Whites. If an individual was needed, but there were questions about his reliability, the Bolsheviks always had the option of threatening not only him but his family. If he betrayed the revolution, they could and would execute his family—a threat that military specialists took very seriously.

The second step taken by the Bolsheviks was the introduction of political commissars. Before we continue, it should be noted that although the term *political commissars* was taken from the French Revolution, the Soviet idea was somewhat different. In the French case, the Defense Ministry had its own officers on the ground. These officers were responsible not only for watching the behavior of line officers, but for morale, motivation, and political socialization as well. The real power, however, resided in the hands of the representatives on mission.

In the Russian case, power was in the hands of the military commissars. It is true that Moscow would send out inspection teams (i.e., senior party officials, who on occasion would report local military and party officials for violating some aspect of party or military policy), but in day-to-day actions, the military commissar was in charge. Another difference, especially when compared to the representatives on mission, was that the political commissar was considered a full-time military officer. He was not just a civilian, out checking on the troops. When it came to combat, the political commissar was expected to be in the midst of the battle, urging his troops on—in this case, like the chaplain during the English Civil War.

As originally conceived, the primary task of the political commissar was to supervise military specialists. In theory, political commissars were not supposed to get involved in technical military matters. Their primary task, in dealing with line officers, was to make certain that the latter did not commit treason. Thus, compared to the French Revolution, the Russian approach was much more systematic, if limited.

In an effort to institutionalize this procedure, the Soviets came up with a policy called "dual command." According to this policy, an order was valid only if it was signed by both the political commissar and the commander. This meant that unlike other military structures, power was divided. To most military officers around the world, such a policy was pure heresy. No longer could the commander rely on his rank and position to issue orders. Before he could give orders, he had to make sure that the political commissar agreed. This had the advantage of ensuring that he never acted against the interests of the party. The commissars themselves were subordinated to higher level military-political organs, which were staffed by more senior political commissars. Indeed, this structure ran all the way from the unit level through larger entities to the top—the Party's Central Committee.

In practical terms, this meant that line officers at all levels were closely supervised. Just as a company commander would have a political commissar working with him, so would a regimental or division commander. In addition, political structures at various levels supervised the actions of the political commissars, making certain that commissars did not permit the commander to issue orders that were not in the party's interest.

In practice, this arrangement created a number of problems. The most obvious was operational. Suppose, for example, that the commander needed to order his troops to make a quick maneuver to block an action by the enemy. Assuming that the political commissar agreed and was right at the commander's side, the issue was relatively simple. If the commissar was not in the area and the commander went ahead and issued the order, he ran the risk of being second-guessed by the commissar when the latter returned—especially if the action was unsuccessful.

Even if the commissar was present, however, the situation could get very difficult. What if the political commissar did not agree? On the one hand, the issue might seem clear—after all, the commander was responsible for operational matters. But let us suppose that the commander proposed that the unit take defensive action, while the commissar, under the influence of the Bolsheviks, who believed they had discovered a new type of doctrine (maneuver and the offense), might insist that the defense was a "defeatist" doctrine and that, as Bolsheviks, they should be attacking.[16] Anyone who has spent time in the military knows that in such a circumstance, time is of the essence. Seconds lost can very easily equal lives lost. In this regard it was a very cumbersome and inefficient form of command. It might ensure that no politically unacceptable orders were issued, but it could also lead to confusion and defeat in battle. In practice, it was often a recipe for inaction. Commanders were afraid to act or, at a minimum, fearful of taking bold steps—unless, of course, they had a very good and healthy relationship with their political commissars.

The dual command structure also had another disadvantageous aspect. While in some cases the two officers worked well together, in other instances the dual command structure led to conflict. For example, military specialists deeply resented the restrictions on their actions that the commissar system imposed. Many of them believed that they had proved their loyalty on the battlefield, where they had shown great courage and devotion to the new Bolshevik regime. Why should they have to clear their actions with a militarily illiterate political commissar? What did he know about military matters? As far as many military specialists were concerned, the answer was simple—very little. If the political commissar wanted to motivate the troops and work on changing their value structure—fine, but stay out of the way of an officer who was trying to win the war. Indeed, throughout the civil war, most line officers longed for the day when these "political hacks" would be removed and sent to teach in party schools or to work at party jobs—anywhere so long as it was not in their present capacity, which many line officers believed served only to advance the interests of the enemy.

For their part, many commissars resented the line officers. After all, most commissars were dedicated party members, individuals who had sacrificed long and hard for the Bolshevik ideal. While they understood the need for military specialists, they failed to comprehend why military specialists should be given such power. From the commissars' perspective, many officers were reactionaries—serving the new regime only because they had no choice. If fortunes were to turn against the Bolsheviks, these officers would be the first to defect to the other side. Furthermore, commissars had seen the face of battle and believed they understood the basics of military science as time went on; thus, they should be given the authority that goes with such knowledge. Besides, increasing numbers of them were discovering that political work at the unit level could be extremely boring and unrewarding. They wanted part of the glory, to be commanders themselves, so that they could lead troops into battle.

Another problem with this arrangement was that it put the average soldier in an impossible situation. What was he to do if in the heat of battle, he received two contradictory orders? Which one should he obey? This was an army with iron discipline, and to disobey an order meant to risk a bullet in the back of the neck. Needless to say, such a situation also created a sense of hesitation in the minds of many soldiers. Better to avoid taking any action that would alienate either officer than to run the risk of getting into trouble and finding oneself facing a firing squad.

Political indoctrination also became a matter of some concern. Over time, the Red Army relied more and more on conscripts. As that happened, the pressure on political commissars and their assistants to carry out political indoctrination lectures expanded considerably. If the Kremlin was going to win this civil war, then it first had to win the hearts and minds of its own soldiers because in many cases they would be fighting against relatives and friends. Furthermore, not only did these soldiers have to be taught the rudiments of Bolshevism, most of them could not read and write—which meant that literacy classes also had to be held. This also placed the commander in conflict with the commissar. Troops needed to train, train, and train some more if they were going to be successful in battle. Yet the political commissar would always be asking for time out for political indoctrination discussions. Every hour spent studying Marx meant one less hour available to learn how to shoot, ride, or repair needed equipment.

Finally, in order to make this system work, both officers found themselves together almost all of the time. "The commissars roomed with the commanders to whom they were attached. They accompanied them wherever they went, so that each step made by the commanders was immediately known to the commissars."[17] Not surprisingly, many commanders resented what they considered to be constant "spying" by the commissars. To make matters worse, the commissars had been declared to be "the direct representative of Soviet power and, as such, to be an inviolable person." Thus, while in theory both were equal, the commissar was often a bit more equal than the commander. As von Hagen put it, "Any insult or other act of violence against a commissar while he was executing his official responsibilities was equivalent to the 'most serious crime' against the Soviet regime."[18] All a line officer would have to do was criticize the actions of a commissar to find himself in hot water. In the eyes of many in the new regime, to criticize a political commissar was to criticize the party he represented.

Lest the reader get the wrong impression, this battle over the appropriate role to be played by the commissar and the commander was no secret. It was openly discussed at party meetings and conferences and by leading party officials. It was one of the most important issues in the military at this time. What more important question was there than the form of command to be utilized in the armed forces? Everyone recognized that political reliability was crucial, but was the party's obsession worthwhile, if it led to the kinds of problems noted previously? How could the military hope to execute this or any other war if the hands of the commander were tied? Something had to be done.

The commissar-commander question came to a head in March 1919. By this point, the positions of proponents on both sides had become polarized. To quote one participant, "There are two points of view; to utilize military specialists and military technicians socialized under capitalism or to get rid of specialists, and in military affairs utilize only those individuals who are completely sympathetic to us."[19] Both positions were clearly argued at the Eighth Party Congress, held from March 18 to 23. Some maintained that the issue had been "historically decided." Like it or not, there were tens of thousands of military specialists in the armed forces. These men had done an outstanding job and they were indispensable. Leaders like Leon Trotsky pointed this out over and over again.

> The broad public knows almost all of the cases of betrayal and treachery by military specialists, but unfortunately, not only the broad public but even more close party circles know very little about all of those military specialists who honestly and consciously died for the cause of the Workers and Peasants of Russia.[20]

Those who supported this point were on solid ground. The young Red Army could not have functioned without these officers. As one person put it at the Congress, "It would be impossible to substitute our Communist officials for military specialists."[21]

The other side was just as fervent in presenting its views. Vladimir Smirnov, a party official who led the attack on military specialists, argued that most of them were secretly sympathetic to the Whites. Instead of giving military specialists more power, he argued that the party should strengthen further the position of the political commissars. Indeed, he went so far as to argue that responsibility for military operations should not be in the hands of the commander, but that it should be invested in revolutionary councils. The only role he would accept for the military specialists was that of an adviser. In his eyes, the principle of "unity of command," which meant that primary power would be in the hands of the commander, would have been a disaster.[22] Others felt the same way. "Some members of the Revolutionary Military Council and in the army considered the principle of unity of command in the leadership of military operations, unacceptable, 'disastrous' for the revolution and insisted on retaining the dual form of command."[23]

To those opposed to giving authority to commanders and line officers, the keys to reliability were party membership and class background. If these were missing, the issue of political reliability would be a moot question. Regardless of what such individuals might say, their capitalist past and support for the Imperial regime made them unacceptable to the new party and its government. In this as in so many other areas, Lenin emerged as a great pragmatist, one who was always prepared to sacrifice ideological purity on the altar of practical expediency. He understood better than many of his more ideological supporters that unless the government maintained power, all of their ideological arguments would count for nothing. This was why he argued, "when you propose theses which are completely directed against the military specialists, you violate the entire tactical position of the party. This is the basis of the difference of opinion."[24]

Unfortunately, Soviet leaders were in a difficult position. General Kolchak's army was moving on Moscow and there were deep splits at the Congress over the peasant question. The last thing leaders needed was another divisive issue. They did not want to do anything that would further antagonize the left wing of the party. As a consequence, the Congress made a concession to the left wing (popularly known as the "military opposition"). It passed a resolution that strengthened the hands of the political commissars by making them representatives of both the Soviet state and the Communist Party.

Ironically, by giving political commissars the added status of state representatives, Soviet authorities were laying the groundwork for a long-term change in the role of the political commissar. Many years down the line, both the individuals carrying out political work and those serving with the political commissar would gradually begin to perceive him to be first a military officer (i.e., a state representative) and only second a political representative.

In taking the action noted previously, the Congress also gave political commissars the right to inflict summary punishment, and the dual authority arrangement was expanded to include such nonpolitical items as administration and supply. Now both officers had to sign not only operational orders but any that dealt with logistics or administrative issues as well. The Congress also called for an intensification of political work and the development of a program that would enable ordinary soldiers to become officers.[25] At the same time, it is worth noting that the pragmatists at the Congress were able to block the ideologues from going as far as they wanted. The pragmatists continually emphasized the positive and crucial role played by the military specialists. As one speaker put it, "Whatever area you take, supply, technical, communications, artillery—for that we need military specialists, because we don't have them."[26] The resolution adopted by the Congress also emphasized the importance of the role played by the military specialists and commended them for their loyalty to the Soviet state. Despite this nice-sounding rhetoric, however, the fact remained that the political commissar's power and influence had been strengthened, making it ever more difficult for the line officer to carry out his duties. Flexibility is one of the keys of command, and for practical purposes, the military specialist's freedom of action had been further restricted.

Because of the increasingly negative impact that this relationship had on the army's command relationship, the issue refused to go away. It was undermining military operations. Those who favored military specialists had a simple solution. For example, by the end of 1919, Ivan Smilga, the head of the Main Political Administration (MPA—the organization that supervised the activities of all political commissars), raised the issue, maintaining that commanders who had proved their loyalty should be allowed to function without interference by political commissars. This meant that in units where the commander had shown his loyalty to the Communist Party and the Soviet state, the post of political commissar should be abolished. In terms of controlling the military, Smilga argued that the party organization (which was present throughout the military) and the political organs (which also were present, especially at higher levels) could do the job. While it may sound strange to hear such comments from the head of the MPA, he was taking his orders from Trotsky, who was primarily interested in putting an effective war machine in the field.

Insofar as his public stance was concerned, Trotsky tried to take the middle ground. For example, in December 1919 he gave a speech in which he argued that it would be possible to dispense with political commissars in some cases. "If military specialists are good workers, there is no reason not to trust them completely in a political sense; it is always possible to observe their actions. And that does not have to be done by a commissar."[27] Being very much the diplomat, Trotsky went out of his way to avoid offending those on the left who wanted to keep political commissars. He made it clear that while he favored the principle of unity of command, he was not arguing for the abolishment of commissars. In reality, however, political commissars were gradually becoming dispensable. As Fedotoff-White put it, "As soon as the Red Army had become a regularly organized armed force, commanded by men welded into it by the fire of battle, the commissar had no place in it."[28]

Despite Trotsky's suggestion that political commissars were increasingly unnecessary, those on the left were not about to give up the fight. They did not trust soldiers of any kind, and most of all they remained suspicious of soldiers who had served the old regime, regardless of how loyal they had been during the civil war. In the minds of left-wing critics, discussion about abolishing political commissars should cease until more Red Commanders had been trained. Not only were political commissars perceived to be more reliable by those on the left, there was a desperate need for these officers. To quote one Soviet source, "Success in combat demands that in the position of a commander there be real commanders . . . and the supply of commanders from military specialists has almost run out. . . . In one division on the Southern Front five regimental commanders were wounded or died the other day." As more Red Commanders entered the military, military specialists would gradually become less necessary. And there is no doubt that the Soviets were making a major effort in this regard. This is evident from the following figures. By 1920 the number of training courses stood at 153. Of this number, 64 were devoted to training officers to deal with machine guns, 17 were for cavalry officers, 8 trained engineers, and the remaining 45 were for a variety of specialties. This net of military schools served some 81,000 individuals.[29]

By the end of the civil war, the Red Army was made up of 130,000 officers. They could be divided into three groups: Red Commanders, NCOs and warrant officers from the old military, and military specialists. Although everyone would have preferred to place primary emphasis on the Red Commanders, the fact was that despite the party's Herculean efforts, training a new officer corps took time. Many of those who wanted to be officers were illiterate. Thus, not only did they have to go through a military school (which included learning their specialty), many of them had to learn the basics of reading, writing, and mathematics. In the meantime, the military specialists would continue to be indispensable.

Training Red Commanders

During the 1920s, the Red Army went through significant force reductions. This gave the Kremlin an opportunity to focus more on quality. A lot of the military specialists could be dispensed with, while primary focus was placed on the Red Commanders. Thus, instead of the 130,000 officers at the end of the civil war, by 1921 there were only 48,000. Of this number, 14,000 were Red Commanders.[30] It is important to emphasize that despite the increase in Red commanders, when it came to quality, they remained hopelessly outclassed by the military specialists. The Red Commanders' training was shorter in duration and their experience on the battlefield—in spite of the civil war—paled in comparison with that of the line officers from the Imperial Army.

Given the senior officers' bias in favor of the more competent, technically qualified officers, it should come as no surprise that the downsizing hurt Red Commanders the most. The generals clearly opted in favor of the better qualified military specialists. In fact, Red Commanders were often singled out and relieved of their command positions because of their "lack of theoretical knowledge."[31] This situation became a matter of concern at the Tenth Party Congress, which was held from March 6 to 16, 1921.

The Congress came out in favor of the Red Commanders. It called for an immediate halt to the practice of releasing communists from the military just because of their lack of military qualifications. The resolution, which the Congress adopted, called for those who had military rank to be reinstated. In addition, it ordered the military to pay more attention to "a planned and systematic use of Red Commanders in command positions."[32] As was often the case when the political leadership decided there was a need to mobilize resources in an area like this, the Congress called for party organizations around the country to become involved. In particular, the Congress argued for "increased attention by soviets, party and professional institutions and organizations to the all-around improvement of the conditions of military-educational affairs."[33]

The Central Committee followed up a month later by calling on local party organizations to assist local educational institutions in recruiting politically active young people and by helping to improve living standards at military schools. This led to the introduction of an educational reform in 1921.

The 1921 reform was aimed at educating a new generation of politically reliable officers who would become the backbone of the Red Army. It was also intended to provide the Red Commanders with the basic knowledge they needed to be competitive with the military specialists. By giving them the background they needed to deal effectively with military matters, the sponsors of the 1921 reform hoped that in time they would be able to replace all of the military specialists.

To begin with, short-term classes were ended. This would permit the educational system to focus attention on longer-term training, the kind that both Red Commanders and young officer candidates needed. Attendance at most schools was lengthened to three or four years so that students could take a broader list of courses and spend more time on the important ones. Nowhere was the increased emphasis on education more evident than at the General Staff Academy, as the energetic and talented Mikhail Tukhachevskiy took over as its commander. It soon became "the leading center for military education." In 1921 Moscow created a Military-Political Academy (to train political officers, especially at the more senior ranks). The next year an air force academy was established. By 1923, there were "55 regular schools, 13 commander's courses, and 10 higher military educational institutions."[34]

The battle between Trotsky and Stalin for power could not avoid having a major impact on the question of political commissars and military specialists. For example, in 1923 a commission was set up to investigate the status of the Red Army. In its report, the commission sharply criticized the quality of the officer corps. The report claimed that the poor level of education in the Red Army was due to the poor qualifications of the instructors (most of whom were military specialists). It went on to blast the political and military qualifications of all officers.

"From a social standpoint up to 45 percent of the officers do not satisfy the demands of the Red Army (former White Officers make up to 53 percent of officers)." The report then noted that "a third have no combat experience, and up to 12 percent lack a military education."[35]

Those who opposed the use of military specialists seized on the report to intensify their attack on the Kremlin's continued reliance on such officers. One opponent criticized the makeup of the officer corps by noting that "In all of our main directorates we find that the domination of old military specialists, and generals has significantly increased." He then read a letter from another critic, which blasted the military leadership for having adopted the spirit of "the old Tsarist military specialists."[36] While most of the attacks on the military leadership at this time were primarily aimed at undermining Trotsky's position, it was also clear that the continued reliance on military specialists to serve in the Red Army was a matter of concern to a lot of party people. In fact, the commission's report clearly called for a reduction in the use of such individuals.

> The commission pointed to the need . . . to keep in the army only those specialists from the former Tsarist army who, as a result of their actions during the Civil War, showed their commitment to Soviet power; it also called for a review of the professors and instructors on the staffs of academies and at other military-educational institutions.[37]

It was becoming increasingly clear that in the future, the number of military specialists would be seriously limited in the Red Army. The majority of line officers would be either Red Commanders or committed young men prepared to master modern military technology. Andrei Bubnov, who was head of the Political Administration of the army, described this policy as aimed at

> the regrouping of the personnel of the military organs to permit a more rapid advancement of the young military communists and of the young "Red Commander" element trained in Soviet military schools or promoted to command positions from the ranks during the Civil War.[38]

By 1924 it was becoming clear that further steps would be necessary to bring the educational level of officers up to the necessary requirements. To begin with, the different services required different lengths of service (infantry was a year and a half, while the navy was four and a half years). Furthermore, the one-and-a-half-year obligation, together with the twice-a-year call-up, caused both economic and training difficulties. It also became clear that a year and a half was not sufficient time to train Russian minorities, most of whom not only did not speak Russian but were closely tied to a much different set of customs and attitudes. The result was a government directive that standardized military service at two years. The only exceptions were the more technical navy and a few other specialties. For the first time in its short history, the Soviet Union now had a standardized policy for dealing with conscripts.

Educational qualifications of officers continued to occupy front stage within the Kremlin. And the situation was not encouraging. By 1924 educational standards were still abysmal. Fifty-six percent of officers had received their military training in the Imperial Army, 12 percent did not have any military education at all, while only 37 percent had completed Soviet military educational institutions.[39]

As a result of a major push on the part of Soviet authorities to increase party membership in the army, by September, the percentage of communists in the central military apparatus increased from 12 to 25 percent. Needless to say, this strengthened the hand of the party stalwarts within the military. At the same time, the percentage of those older than fifty dropped from 21 to 5 percent. Since most of those over fifty were military specialists, this hit them the hardest.[40]

Meanwhile, progress was made in training Red Commanders. While in 1922 the percentage of Red Commanders in the officer corps stood at only 22.5 percent, by the following year it had risen to 29.5 percent. By 1924 it was up to 31.8 percent.[41] At the same time, the number of officers who had graduated from Soviet military academies increased ten times, as the size of the Red Army was reduced from 5 million to just under 652,000.[42]

In an effort to leave nothing to chance, the Kremlin issued a decree on November 26, 1924, entitled "A Network of Military-Educational Institutions." This decree stipulated that the best commanders were to be sent to teach at military educational institutions. It also ordered that minimal standards for students be raised and that, while cadets, students would serve part time with active duty units. As a result, the number of instructors was increased from 21,348 on November 1, 1924, to 24,028 a year later. Officers were also given the option of taking short-term courses in a variety of technical areas. The numbers of graduates indicate that the Soviets were making progress. "In 1924 military educational establishments produced 6,848 commanders, in 1925, 9,193 graduates. Besides, during 1925, 258 individuals graduated from military academies." By the end of 1926 the percentage of officers with a military education stood at 95.5 percent.[43]

At the same time that Moscow was working to produce a new generation of military officers, a purge was under way among military specialists. By January 1, 1925, the number of military specialists on the main army staff had been reduced from 2,598 (June 1, 1923) to 397. Speaking about those who were retained, one source emphasized their political reliability, remarking that "the majority of them had shown their devotion to the working class."[44]

Political criteria continued to play a key role among officers. For example, the percentage of officers having a working-class background increased from 13.6 percent on June 1, 1923, to 20 percent on January 1, 1925, while the percentage of party members rose from 20 percent to around 40.9 percent during the same time period. At the same time, party membership and appropriate class background were playing a major role for those who wanted to attend officer's schools. To cite figures from the same source: the graduating classes of 1925 were made up of

33 percent workers, 52 percent peasants, and 15 percent others, while 51.4 percent were communists, 15.5 percent were members of Komsomol, and only 33.1 percent were not party members.[45] By the end of 1925 the Kremlin could argue that it had succeeded in

> wiping out the differences between the Red Commanders and the military specialists; between officers having finished Soviet educational institutions—the Red Commanders and the officers of the old army, who as a result of their honest and selfless service in the Red Army, have demonstrated their loyalty to Soviet power.[46]

The time for modifying the relationship between commanders and political commissars was at hand.

Given the political baggage that surrounded the issue of political commissars, the key political question was how to operationalize a change in status. Should the commissar be dropped in favor of making the commander the key decision maker in all military issues? Or should he be retained, albeit in a weakened or modified form?

First efforts to modify the role played by political commissars occurred in 1922. A number of commanders who had completed military academies and higher military schools, and who had been members of the Communist Party for at least two years, were allowed to combine both functions. A political assistant took the place of the political commissar in such cases. The political assistant retained the right to pass on personnel changes and to write a political fitness report on the commander—as well as on all other officers he served with. The latter authority was especially important because it "gave him a very substantial role in all promotions of commanders by means of personnel files in which he recorded the commanders' work habits, political activities, and other information that entered into consideration for promotion."[47] He was also responsible to the commander for things like motivation, morale, and political socialization. While very few line officers enjoyed this kind of freedom of action in practice, it did set a precedent.

Mikhail Frunze, a man who was to play the key role in the area of major reform in the Red Army in the mid-1920s, began to argue that the conditions for a transfer to the "one-man command" system now existed. Bubnov made the same point at a meeting in June when he observed, "We have in the ranks of the Red Army in sufficient measure selected, firm and in all areas qualified officers." Stalin took a similar line. In the end, the commission adopted a resolution, noting that it "recognizes that unity of command is the most efficient principle for the development of the Red Army."[48] Regarding the implementation of this new idea, the commission left the details up to the military leadership.

The functions that political commissars should carry out were again addressed in December when the Revolutionary Military Council passed a resolution that called upon commissars to restrict their actions to ensuring the "moral-political condition" of the unit in which they served. Then a directive, called "On the Involvement of Officers in Political Work," was issued. This directive ordered all officers—both political and line—to take part in the political education of the troops serving under them. The directive made clear that all officers would be evaluated based on their performance in this area. The same directive also noted that since some officers still did not accept fully the new communist regime, in some units the unity of command principle would be only partially implemented. Frunze elaborated on this idea in a speech he gave at the same time. According to him, in the immediate future, there would be three types of command in the Red Army. In the first case, line officers would be responsible not only for combat and administrative/logistical matters but for political work as well. While this was the goal of the new system, only a few officers would be given such broad responsibility in the beginning. The second category put the commander in charge of administrative and logistical matters, as well as combat. He would not be in charge of political work—and this was the most common form of command in the Red Army at this time. Finally, a third type would exist in administrative organs such as staffs and directorates. A month later, Frunze used his authority as commissar for war to abolish the titles of "military specialist" and "Red Commander." All individuals, regardless of their function, would now carry the more simple title of "officer."[49]

In spite of the revolutionary nature of Frunze's actions, there were still complaints. In particular, certain line officers believed that the army should get rid of political commissars and political officers altogether. As Frunze himself noted in 1925, "Some commanders underestimate and minimize the importance of political work in the Red Army in general."[50]

The political commissars and their supporters fought back. They argued that the unity of command principle should only be adopted if the officers were Communists. Officers who were not party members should only be permitted to exercise the unity of command principle in exceptional cases—and then not in operational commands but as members of staffs.

> Unity of command cannot be viewed as a liquidation of the institute of commissars, because it is not possible to liquidate elements of the party leadership and political education in the army. Unity of command must be viewed as the joint function of party leadership, political education, military training in one person—the single commander.[51]

Unity of command was again discussed at a plenum of the Communist Party, which was held at the end of November and the beginning of December 1924. The plenum argued in favor of increasing the limited autonomy enjoyed by regular officers. It decided that unity of command should be introduced in two forms. The first was the so-called "incomplete" unity of command. This was obviously aimed at those individuals whom the political leadership did not consider to be ready to assume complete control over all aspects of military life. In this instance, the commissar would continue to be charge of political matters. In order for an order to be valid, it still had to be signed by both officers. The second or "complete" form of unity of command gave the commander total freedom in issuing orders. The political commissar became his political assistant. This approach was formalized with the issuance of Directive 234 in March 1925.[52]

In reality, progress toward the unity of command principle was slow. Of 44,326 officers in command positions as of October 1, 1925, only 1,184 of them (2.67 percent) were appointed unified commanders. This was most notable among middle- and lower-level commanders. Among middle-level commanders, for example, only .083 percent were given that title.[53] The political commissars were still very much in charge. On the other hand, the precedent had been set. Once line officers could prove they were reliable, they could be placed in a unity of command position. Besides, anyone with any sense of how bureaucracies work knew that it was only a matter of time before the change in relationship would become the norm. An organization like the military demands standardization and cannot work with two different systems of control over the long run.

The change that had occurred was particularly evident if one compared the new directive with the one that had been issued on January 8, 1922—the one that had been in force up to that time. The 1922 directive stipulated that the role of the political commissar was to carry out "political control and the direct observation of line officers." The new directive spoke of the commissar's role as "leading and the immediate supervision of party-political work and ensuring the training and education of personnel of the Red Army and the Red Fleet in a spirit of class unity."[54] The latter language focused his work much more on areas such as motivation, morale, and political socialization than it did on control. This is not to suggest that the political commissar did not have responsibility for these functions; he did. But now he would be able to direct his attention more firmly in these areas.

The implication of this move was clear. Once officers were sufficiently trained and loyal, the full form of Frunze's idea of unity of command would be introduced. The writing was on the wall insofar as the commissars were concerned, and they did not like the message at all. Frunze touched on this when he wrote that "it is impossible not to mention that in some units our commissars continue to act negatively toward all of these reforms."[55] As an effort to meet the political commissars half way, the Kremlin launched a campaign offering commissars appointments as line officers—assuming, of course, that they had the necessary technical qualifications. In 1925 a certification board reviewed the files of every political commissar serving in the armed forces. The board decided to make a large number of them line officers and military commanders.

Progress toward the unity of command principle was soon evident. For example, by September 1926 100 percent of corps commanders, 54.7 percent of division commanders, 36.5 percent of regimental commanders, 37.7 percent of company commanders, and 75.5 percent of the chiefs of military educational institutes operated under the unity of command principle.[56]

The next important step came in 1927. Klement Voroshilov, who succeeded Frunze as commissar for war, issued Order No. 11. This order stipulated that a commander who did not have a political commissar (one who was operating under the unity of command principle) could sign any order—except for those dealing directly with political affairs. While his political deputy would remain in charge of political matters, he would work for the line commander. Any disputes between the two officers were to be reported to the next highest political organ. In this capacity, the political deputy was responsible for things like political work, morale, discipline, training, and housekeeping matters. In order to make sure that the two officers worked together, however, the order clearly stated that the two "bear complete responsibility for the political and morale condition of the unit as well as its combat readiness."[57]

By the beginning of 1930 the percentage of regimental commanders exercising unity of command had risen from 48 percent in 1927 to 73 percent. Likewise, the percentage of company commanders enjoying this authority rose from 36 percent in 1927 to 50 percent in 1950.[58] By the end of 1931, the transition to the unity of command system was almost complete. The overwhelming majority of units operated under it. The only exceptions were the navy and the air force, where the "party stratum" included fewer party members. In the vast majority of cases the commander's authority remained limited until May 1937, when the dual command system was reintroduced.

Conclusion

Political commissars functioned in a somewhat different manner than did their ideological predecessors during the French Revolution. Instead of being civilians who were sent out on missions to ensure the military's loyalty, they were part and parcel of the Red Army. They wore the same uniform, ate the same food, slept in the same tents, and were subject to the same orders as all other officers. In this sense, they were more like the military commissars who served in the French Revolution, except that their power over commanders and line officers was greater.

What was most significant about them, however, was how their role changed drastically over time. Undoubtedly, they were primarily a control device when they were first introduced. The Bolsheviks feared the military specialists (often with good reason) and so had to come up with a vehicle for making certain that they obeyed the orders they were given. Commissars were very useful in this regard.

There was a trade-off, however. By utilizing commissars in this fashion, the military became less efficient as unit cohesion was undermined, and confusion, if not chaos, was often created because two individuals had to agree on almost everything that the unit was to do before the order became effective.

In time, however, the situation began to change. The commissars functioned less as control devices. The "foreign" specialists that the party had to rely upon were fewer in number. Some, like Mikhail Tuchaschevskiy and Boris Shaposhnikov, would go on to play major roles in the evolution of the Red Army. But for practical purposes, the military specialists became unnecessary as the effort to train younger and more politically reliable line officers was increasingly successful. Having reached the point where there was no need to assume that line officers were unreliable, the party decided to modify this militarily inefficient control device by introducing the political assistant.

The change in name from political commissar to political assistant involved far more than just a modification in the officer's title. The commander became the unit's commander in fact, as well as title. He would issue orders on his authority alone—except in the more narrow political sense, where the political officer worked under the direction of the Main Political Administration.

Does this mean that the political commissar had been neutered? That his influence in the chain of command had come to an end? Not at all. The political assistant still filled out reports on line officers and if he decided to, he could make their lives very difficult. All that was required was a report noting that so and so was not carrying out his obligations as a good party member—or the political assistant could bring up the issue within the party organization to which almost all officers would eventually belong. In either case, the line officer would be in serious trouble. The political assistant also continued to lead the charge in the political area—working to socialize the mass of soldiers and officers into the new Bolshevik belief system.

One lesson that both officers soon learned was that since they shared responsibility for how the unit performed, working together was a much more effective way to get the job done. This does not mean that the two of them always agreed. After all, political officers inevitably called for political indoctrination lectures or meetings at times when the commander or other line officers wanted the troops out training. Both officers learned, however, that if they wanted to survive, they had to find a way to get the two jobs done. A unit that scored high on political tests but that failed military exams was in bad shape, and the commander (as well as the political assistant) could expect to be called on the carpet to explain why performance was at such a low level.

If there are any lessons in the evolution of the role played by political commissars in the Soviet armed forces, they are twofold. First, it is clear that when push came to shove, successful revolutionaries tended to be pragmatic. This was certainly the case with Lenin, Trotsky, and Frunze. They were more than prepared to take on the ideologues when the survival of the regime was at stake. Though all three of them would undoubtedly have been much happier if the Red Army had been composed entirely of politically reliable and technically competent communists, that was an impossibility. So they did the next best thing: they took the institution of political commissars that had existed during the French Revolution and modified it to fit Soviet circumstances.

Second, it is important to note how this control measure was modified during the ten or so years it was in operation. The Bolshevik leaders were not stupid and realized that they needed to change the system as soon as possible. And it changed.

The position of political assistant that came into being during the early 1930s was quite different from that of the commissar of 1918. The political assistant was still an important part of the Red Army, but his ability to issue orders had been sharply circumvented.

Notes

1. As noted in John Erickson, *The Soviet High Command* (New York: Macmillan, 1962), 7.

2. The following is based in part on chapter 4 of my *Russian Civil-Military Relations* (Bloomington: Indiana University Press, 1996), 56–70.

3. Erickson, *The Soviet High Command*, 16.

4. S. I. Gusev, *Grazhdanskaya voyna i Krasnaya Armiya* (Moscow: Voennoe Izdatel'stvo Ministerstva Oborony Soyuza SSR, 1958), 28.

5. A. G. Kavtaradze, *Voennye spetsialisty na sluzhbe Respubliki Sovetov, 1917–1920gg* (Moscow: Nauka, 1988), 74.

6. N. I. Shatagin, *Organizatsiya i stroytel'stvo Sovetskoy Armii v period inostrannoy voennoy interventsii i grazhdanskoy voyny (1918–1920)* (Moscow: Voennoe Izdatel'stvo, 1954), 52.

7. N. A. Kozlov, *Adademiya general'nogo shtaba* (Moscow: Voennoe Izdatel'stvo, 1987), 11–12.

8. Kavtaradze, *Voennye spetsialisty na sluzhbe Respubliki Sovetov*, 70, 198.

9. L. I. Trotsky, "Nam nuzhna armiya," in L. Trotsky, *Kak zooruzhalas' revolyutsiya* (Moscow: Vysshiy Redaktsionnyy Sovet, 1923), vol. 1, 27–28.

10. Kavtaradze, *Voennye spetsialisty na sluzhbe Respubliki Sovetov*, 77.

11. A. F. Danilevskiy, *V. I. Lenin i voprosy voennogo stroytel'stva na VIII S"ezde RKP(b)* (Moscow: Voennoe Izdatel'stvo, 1964), 30.

12. As cited in Kavtaradze, *Voennye spetsialisty na sluzhbe Respubliki Sovetov*, 95, 96, 97.

13. Shatagin, *Organizatsiya i stroytel'stvo Sovetskoy Armii*, 52.

14. Kavtaradze, *Voennye spetsialisty na sluzhbe Respublki Sovetov*, 207–209.

15. Kavtaradze, *Voennye spetsialisty na sluzhbe Respubliki Sovetov*, 106.

16. The debate over military doctrine is discussed in chapter 1 of my *Russian Civil-Military Relations*, 3–20.

17. Dmitri Fedotoff-White, *The Growth of the Red Army* (Princeton: Princeton University Press, 1944), 84.

18. Mark Von Hagen, *Soldiers in the Proletarian Dictatorship*, (Ithaca, N.Y.: Cornell University Press, 1990) 32.

19. As cited in Kavtaradze, *Voennye spetsialisty na sluzhbe Respubliki Sovetov*, 121.

20. L. Trotsky, "Voennye spetsialisty i Krasnaya Armiya," *Voennoe Delo* 2 (1919): 71.

21. *Vos'moy s"ezd RKP(b)* (Moscow: Partiynoe Izdatel'stvo, 1933), 148–149.

22. *Vos'moy s"ezd RKP(b)*, 155–160.

23. *Vos'moy s"ezd RKP(b)*, 155–160.

24. *Vos'moy S"ezd RKP(b)*, 539.

25. *KPS v rezolyutsiyakh i resheniyakh*, Part I (Moscow: Gosudarstvennoe Izdatel'stvo Politicheskoy Literatury, 1954), 439–440.

26. Danilevskiy, *V. I. Lenin i voprosy voennogo stroytel'stva na VIII S"ezde RKP(b)*, 76.

27. L. Trotsky, "Nashi ocherednye voprosy," in L. Trotsky, *Kak vooruzhalas' revolyutiya*, vol. 2, book 1 (Moscow: Vysshiy Voennyy Redaksionnnyy Sovet, 1924), 77.

28. Fedotoff-White, *The Growth of the Red Army,* 85, 89–90.

29. I. V., "Nam nuzhen komandnyy sostav," *Voennoe delo* 2, no. 3 (1919): 157.

30. K. E. Voroshilov, "Desyatiletie Krasnoy Armii," in K. E. Voroshilov, *Stat'i i rechi* (Moscow: Partizdat TsK VKP(b), 1936), 228.

31. I. B. Berkhin, *Voennaya reforma v SSSR* (Moscow: Voennoe Izdatel'stvo, 1958), 38.

32. "Desyatyy s"ezd RKP(b)," in *KPSS v rezolyutsiyakh i resheniyakh s'ezdov* . . . vol. 1, 568–571.

33. "O komplektovanii voenno-uchebnykh zavedeniy," in *KPSS o vooruzhennykh silakh Sovetskogo Soyuza* (Moscow: Voennoe Izdatel'stvo Ministerstrva Oborony SSSR, 1969), 166–167.

34. A. Yovlev, "Sovershenstvovanie voenno-uchebnykh zavedeniy v 1921–1928gg," *Voenno-istoricheskiy Zhurnal* 2 (1979): 95.

35. As cited in Berkhin, *Voennaya reforma v SSSR,* 68.

36. As cited in Berkhin, *Voennaya reforma v SSSR,* 61–62.

37. Berkhin, *Voennaya reforma v SSSR,* 68.

38. As cited in Fedotoff-White, *The Growth of the Red Army,* 199.

39. Berkhin, *Voennaya reforma v SSSR,* 259.

40. Berkhin, *Voennaya reforma v SSSR,* 155.

41. Petrov, *Stroytel'stvo politorganov, partiynikh i komsomol'skikh organizatsiy armii i flota,* (Moscow: Voennoe Izdatel'stvo Ministersva Oborony SSSR, 1968) 156.

42. Berkhin, *Voennaya reforma v SSSR,* 77.

43. Yovlev, "Sovershenstvovanie voenno-uchebnykh zavedeniy," 95.

44. Berkhin, *Voennaya reforma v SSSR,* 80.

45. Berkhin, *Voennaya reforma v SSSR,* 262, 273

46. Berkhin, *Voennaya reforma v SSSR,* 263.

47. See von Hagen, *Soldiers in the Proletarian Dictatorship,* 163–164.

48. A. Bubnov, *1924 god v voennom stroytel'stve* (Moscow: Voenizdat, 1925), 87.

49. M. V. Frunze, "Itogi plenuma Revvoensoveta SSSR," in Frunze, *Izbrannye proizvedeniya,* 265–266.

50. M. V. Frunze, "Ochernye zadachi politrabotnikov," *Voyna i revolyutsiya* 5 (August–September 1925): 5.

51. Berkhin, *Voennaya reforma v SSSR,* 294.

52. "Ob edinonachalii v Krasnoy Armii," in *KPSS o vooruzhennyh silakh Sovetskogo Soyuza* (Moscow: Voennoe Izdatel'stvo Ministerstva Oborony SSSR, 1969), 228–229.

53. Berkhin, *Voennaya reforma v SSSR,* 306, 395.

54. Berkhin, *Voennaya reforma v SSSR,* 396–397.

55. Frunze, "Ochernye zadachi politrabotnikov," 6.

56. Voroshilov, *Oborona SSSR,* (Moscow: Voenny Vestnik, 1927) 152.

57. Voroshilov, *Oborona SSSR,* 77–78.

58. Yu. Petrov, "Deyatel'nost Kommunisticheskoy partii po predvediyu edinonachaliya v Vorruzhennykh Silakh," *Voenno-istorcheskiy Zhurnal* 5 (1963): 19.

5

National Leadership Officers in the Wehrmacht

Two institutions, the German army and the National Socialist Party found themselves face to face. . . . the question was, who would gain the upper hand in the new German state, the party or the army.

—Andre François-Poncet, French ambassador to Berlin

O F ALL THE RELATIONSHIPS DISCUSSED in this book, that between the party and the military in Hitler's Germany is one of the most interesting and unique and the least known.[1] The German experience was unique because it is the only case in this book where chaplains and political officers existed side by side.[2] The actions of chaplains were very circumscribed, but these men continued to carry out their duties right up until the end of the Third Reich.

The German case is also unique because rather than purging the officer corps—which, at the top, had values very different from those espoused by Hitler and his followers—Hitler left the military alone, at least during his early years of power. The reason was simple. The party was faced with a very autonomous, highly professional military, and Hitler was dependent on it to achieve his foreign policy goals. Without it, he would never be able to enhance German power and Germany's role in the world. Later on, it was the only institution that stood between Berlin and military defeat. As a consequence, he was not in a position to subordinate the military to the party's will, as happened in revolutionary France or in the former USSR. In those instances, the party purged the military of what it considered unreliable elements and placed it under close supervision and control—to ensure that it did not act against the new revolutionary government. The values of the officer corps in both cases were so divergent from those of the new political elite that the regimes felt they had no alternative but to infiltrate and atomize the armed forces.

The situation in Hitler's Germany was quite different. The difference in values was not nearly as great. While some in the armed forces were suspicious of Hitler, many others openly supported him—if not his ideology. From their standpoint, his praise for the German military, as well as his struggle against the restrictions placed on the German Army by the Versailles Treaty, was welcome. The generals saw Hitler's accession to power as a very positive development—if for no other reason than that he promised to rebuild and expand the 100,000-strong Reichswehr. Once the numerical restrictions were gone and a remilitarization program was under way, it would only be a matter of time before Germany again assumed its rightful place in world. The Bavarian corporal might not be a military genius, and many of his ideas sounded strange to Reichswehr officers, but he was a means to an end. By working closely with him, the generals felt that the German armed forces had a real chance of being built up to the point where they would again be taken seriously by major world powers.

From a conceptual standpoint, Germany was somewhere between the French and Russian cases, on one hand, and the English situation, on the other. Like Cromwell's officers, the majority of German officers supported most of the policies advocated by the Führer. There was a problem, however, when it came to internal military matters. Given their desire to control all aspects of life, Nazi Party members wanted to be in charge of all military matters, including things such as personnel, indoctrination, ideology, strategy, and tactics. On the other hand, when one considers how important the military was to Hitler's plans and the fact that most of the officer corps supported him, Hitler realized that any attempt to undermine military structure and cohesion would be counterproductive.

Nevertheless, from the perspective of many of the Nazi Party's senior officials, this was a marriage of convenience at best. The idea of having a military that was not under the party's direct control was unthinkable. Yet they could do little about it. Hitler was not prepared to sit and wait until the armed forces were purged and resocialized to fit the Nazi model before proceeding to use them to achieve his goals. He would eventually introduce the Waffen SS as a second, highly politicized military and a potential rival to the military, but for the present, these forces were to supplement the military, not replace it. As a result, the control devices used in other cases—for example, commissars—were not a realistic option in the German case. The military was as jealous of its bureaucratic prerogatives as the party was interested in penetrating the armed forces.

Not surprisingly, the marriage between the military and the party would be marked by conflict, as the military repeatedly tried to minimize party influence—while willingly carrying out the Führer's orders. For its part, the party would try over and over to increase its influence on the officers and men who served Germany. By the end of the war in 1945, the Nazi Party would have succeeded beyond its wildest dreams. Unfortunately, from its perspective the victory would be a hollow one because by the time the party had successfully infiltrated the military, the latter would be on its last legs. The situation facing senior Nazis in 1945 resembled that in which a general reports that he has conquered his objective by destroying it.

The Army and National Socialism

Before we discuss Hitler's accession to power, it is important to explain how the generals—those who had remained with the Reichswehr after the end of World War I—saw the world of politics. Traditionally, the German military had shunned any involvement in politics. They even had a word for it—*Überparteilichkeit*. This meant that the military would remain "above politics." This apolitical approach to politics went back hundreds of years—the task of the Prussian and later the German Army was not to make policy; rather, it was to implement the policy directives made by the country's leaders. Indeed, German officers took great satisfaction in not having anything to do with politics. Their world was the army, and in a certain sense the German Army was a *Staat im Staat*, a state within a state.

Of all the aspects of military life, none figured more prominently in the life of the German officer than the concept of honor. A German officer's word was as good as a written agreement. The concept was medieval, which meant it was closely bound up with the idea of chivalry. The concept of honor in the pre–World War I army was also aristocratic (many, if not most, of the senior officers came from aristocratic families, and those who did not tried to emulate the manners of the aristocracy). Finally, the army was devoted to the monarchy, and most officers had little understanding of, or patience with the political process. From the perspective of military officers, political parties complicated matters. Following the orders of the sovereign was more orderly and more efficient in the eyes of many soldiers.

The situation changed slightly in the aftermath of the defeat in World War I and the 1918 revolution. If anything, looking at the disorder around them, German officers began to place more emphasis on obedience and order. Their already limited patience with the often chaotic world of political parties deteriorated to the point where many of them had nothing but disdain for parties and the existing political process. They longed for the order of the pre-World War I period.

> The struggle for life and death turned absolute obedience into a supreme and unqualified value, and officers from the old Imperial Army, led by General Hans von Seekt, based their whole thought and feelings as soldiers on a directive from Seekt which based honor on obedience.[3]

This heavy emphasis on honor, obedience, and a disdain for the democratic political process would play a major role when Hindenburg died and the German Army took a personal oath to Hitler. Order would be restored and obedience would be absolute, no questions asked.

Despite their commitment to honor and obedience, the country's military officers also believed that just as it was wrong for them to become involved in the political process, it was equally wrong for the government to get involved in internal military matters. They would willingly do whatever political authorities requested, but the idea that the country's political leadership would interfere in areas such as training, promotions, strategy, or tactics was an anathema to them.

They—not the politicians—were the professionals. And as the Hitler period would demonstrate, military officers would fight tenaciously to maintain that autonomy.

Hitler and the Nazis faced a somewhat divided military when it came to political attitudes. Despite the military's apolitical tradition, some officers (generally junior in rank) went against tradition and openly supported the Nazis as early as 1933. On occasion, this led to their being reprimanded by senior officers who objected to their involvement in politics. The generals' repeated efforts to keep the military apolitical were made more difficult by Hitler's willingness to offer the army something it could not ignore—he not only promised an expanded and modernized Germany, he also advocated the emergence of Germany as the dominant military power in Europe.

Meanwhile, many senior military officers continued to be deeply upset by the chaos and lack of unity that afflicted the country in the post-World War I years. Many of them blamed the confusion on the parties that were fighting for different programs. The Nazi Party seemed like a savior. It offered a romantic vision of a unified people, of heroic deeds committed by superhumans. As one source put it,

> In the minds of a majority of the soldiers of the Reichswehr, there were a whole number of parallels between the military view of the world and the patriotic-community theses of the national-socialist world view. This included the strong national state, the elimination of the Versailles "Diktat," the removal of the many parties, the reappearance of military virtues, and the establishment of executive authority, to name the most important.[4]

To make matters worse, a lot of soldiers believed they had been rejected by German society. They felt that many German citizens believed they were responsible for the mess Germany found itself in. After all, quite a few Germans thought the army had lost the war. The view of the Nazis was just the opposite. Hitler's speeches were full of praise for the army. Indeed, he hardly ever gave a speech in which he did not single out the army for special attention, glorifying it and its virtues—while placing the blame for the loss of World War I on the home front. It is thus not surprising that many in the armed forces were susceptible to Hitler and his appeal. As one writer put it,

> This appearance of complete confidence in his subordinates, coupled with his strong interest in rebuilding Germany's armed strength, must have done much to create an atmosphere of trust. The generals probably felt that while dealing with an Austrian corporal might not be very gratifying, things could be a lot worse.[5]

At the same time, some officers were offended by the proletarian and vulgar nature not only of the party's doctrine but also of those who filled its ranks. For people with aristocratic backgrounds—or for those who tried to emulate such a heritage—the idea of associating with a bunch of Nazi thugs was more than they could bear.

This put many officers in a quandary. What to do? They wanted the benefits that association with Hitler promised, but at the same time, they realized that a marriage with the Nazis could become all-encompassing. The result was a duality in military thinking. They would accept National Socialism as the basis for the new army, but they would do everything possible to limit its influence within the military. In practice, this would mean preserving the Prussian ideals of honor and obedience, while at the same time avoiding involvement in political matters and keeping the Nazi Party from infiltrating the armed forces. To quote one writer,

> Despite such differences of emphasis, army command and armed forces command reacted equally firmly whenever members of the party threatened to encroach on their sphere. In their view, this would have contravened the assumption, fundamental to the military elite's claim to autonomy, that they were partners with equal rights. Thus Reichenau, speaking to officers of the army corps in winter 1935, voiced his opinion of the growing volume of reproaches from the party and SS in terms as forceful as they were characteristic. "We do not need to make the soldier into a National Socialist. . . . We are National Socialists without party cards . . . the best and most serious."[6]

Or as General Beck himself noted, "Quite apart from the fact that the Army's basis today is National Socialist, as it must be, the Party's influence must not be permitted to penetrate the Army, for it could have a destructive effect."[7] Nazis, yes, but soldiers first.

As far as political tactics were concerned, Germany's generals believed they could use National Socialism for their own ends. After all, they were educated, well bred, sophisticated, and from the country's most respected institution. Many of them believed that they could educate Hitler—help him understand the complexities of military affairs, while keeping him and his party at arm's length. Unfortunately, they would learn in time not only that they were politically naive but that Hitler was a lot smarter than many in the armed forces realized. He may have only been a corporal during the First World War. However, his political skills would rocket him to the highest position in the country, with far more power than any of the generals ever imagined.

Hitler Comes to Power

On January 30, 1933, Hitler became chancellor. Much to his chagrin, his control of the army was very limited. Field-Marshal Paul von Hindenburg, the hero of World War I and Germany's president, had appointed Field-Marshal Werner von Blomberg, a man he trusted, to be Minister of Defense. This put Hitler in a difficult situation. If he fired Blomberg, he ran the risk of alienating both Hindenburg and the country's senior military officers because he had made it clear to the generals that he did not intend to interfere in their internal affairs. Furthermore, in the belief that they could control Hitler, these military officers had lobbied for Hitler's appointment as chancellor. As a consequence, he could hardly have fired Blomberg at this point.

As Hitler repeated many times in his speeches, power in Germany rested on two pillars—the Nazi Party and the Wehrmacht. It was Hitler's hope that over time the two would grow increasingly together. He wanted party members to take on the characteristics of the military—discipline, devotion to duty, and obedience to those in superior positions. At the same time, he wanted the military to be as committed to Nazi ideology as the party members were. Total dedication to the Führer and to the party was expected from everyone who wore the German uniform.

When it came to converting the military to Nazism, Hitler believed that more than superficial dedication was required. If the officer corps was not totally committed to the Nazi Party and ideology, if the country's senior officers believed that the party's ideas were something that had to be accepted officially but could be ignored or pushed into the background in practice, the Wehrmacht would never have become the instrument that would be so crucial to Hitler's efforts to expand German power. As O'Neill put it,

> If the party was to have a secure tenure of power and absolute authority it was necessary to convert the army to Nazism. . . . Since the army was cast in so vital a role for the future of the Third Reich, and was to bear the brunt of the coming struggles, its fidelity to Party ideology and to the Führer had to be beyond doubt and capable of enduring the most difficult circumstances.[8]

In this sense, it is important to keep in mind that Hitler placed much more faith in emotion than he did in intellect. He would periodically rail against intellectuals, and when it came time to create National Socialist Leadership Officers, he would emphasize to his subordinates that he did not want intellectuals. Hitler believed that if the army embodied the kind of total dedication that he did, it would be invincible. A strong, totally committed will can make up for many shortcomings—from inferiority in numbers to a lack of proper equipment. When the German Army later faltered in the field, Hitler proclaimed that it was because his generals and officers lacked the necessary will.

The SA

One of the first issues that confronted the military and Hitler was the SA (Sturmabteilung). The SA was made up primarily of Nazi loyalists, many if not most of whom had little or no military training. They were first and foremost party members. Many were unemployed and had spent most of the last several years fighting communists or beating up Jews. Furthermore, many of them had pretensions of becoming part of a major military force—a party army. This was particularly the case for the SA leader Ernst Röhm. His goal was rather simple. He wanted to replace the army with the SA—an ideologically committed military. Not surprisingly, this presented Hitler with a dilemma. He needed the army to consolidate his power and implement his nationalistic policies. At the same time, he owed his position within the German government to those members of the SA who had supported him in the many street battles that had been part and parcel of his rise to power. Without their support and assistance in fighting the communists and intimidating his enemies, he would have never become chancellor.

Meanwhile, the army grew increasingly concerned about the SA. It had increased from 300,000 in the period prior to Hitler's seizure of power to well over 4 million by the beginning of 1934.[9] Ironically, this put the military (which was still limited to 100,000 troops by the Versailles Treaty) in a very difficult situation. How should the army respond to Hitler? The generals needed him to keep the SA from overwhelming the armed forces. Hitler recognized the army's vulnerability and "did not hesitate to exploit this period of weakness on the part of the army to spread Nazism as widely as possible throughout the army, from the bottom upwards, thereby circumscribing any opposition to his regime from within the upper echelons of the army."[10] This was accomplished in a number of ways. For example, on May 17, 1933, it was agreed that when it came to military questions, party organizations such as the SA, the SS, and the Stahlhelm would come under the Defense Ministry. In addition, the SA agreed to limit its armaments to hand weapons, leaving the heavier weapons in the hands of the army. In August, the armed forces and the SA agreed that the SA (and other Nazi-controlled organizations) would be the main source of recruits for the army. In practice, this meant that young men would go through a sort of pre–boot camp period of training and political indoctrination before they joined the military. If nothing else, this would ensure that these young men were familiar with the Nazi Party and all that it stood for when they put on Wehrmacht uniforms.

In the meantime, conflicts between the SA and the army were growing. For example, many army personnel refused to accept the order requiring them to salute SA flags. They did not understand why they should show respect toward a nonmilitary or nonnational flag. When they refused to salute it, fistfights between SA and military personnel were often the result. Indeed, by the beginning of 1934, the military was increasingly convinced that long-term cooperation with the SA would be impossible. No matter what the military did, Röhm and his SA colleagues still had the same goal—to eliminate the army's monopoly over the use of force.

By February 1, 1934, Röhm was arguing that the army should be placed under the control of the SA. He maintained that the army's primary function should be to train soldiers for the SA. The officers were to be turned over to the SA, where they would function as advisers. "Röhm sent the high command a message stating that 'I regard the Reichswehr now only as a training school for the German people. The conduct of war, and therefore of mobilization as well, in future is the task of the SA.'"[11] By mid-February Röhm was pushing for the creation of a single ministry dealing with military matters—with himself as its head. Hitler decided the issue in a speech on February 28, 1934, in which he declared that the "Wehrmacht will be the sole bearer of arms in the nation." To emphasize his point, Röhm was forced to sign an agreement in front of Hitler, acknowledging the superiority of the military in this area.[12]

Despite his having signed the agreement, the military did not believe that Röhm would carry it out. Once again, the military's relationship with Hitler came into play because his ability to achieve his goals continued to depend on the military. The same was true of the army. Good relations with Der Führer were critical. In an effort to cozy up to the Führer, General Blomberg published a directive on May 25 entitled, "The Wehrmacht and National Socialism." The purpose was clearly to convince Hitler that the Wehrmacht was behind him and the Nazi Party. The problem was the SA, not the armed forces.

In early June, Hitler and Röhm had their last conversation. It was clear that the positions of the two men were irreconcilable. Röhm was determined to expand the size and tasks of the SA. Such an action on his part would inevitably lead to conflict with the army.

By the end of June, matters were clearly coming to a head. On June 29, 1934, Blomberg wrote an article in the Nazi newspaper *Völkischer Beobachter*, in which he declared that the policy of *Überparteilichkeit* was now outdated. According to Blomberg, "The soldier now stands in the middle of political life. The Wehrmacht serves this state, which it approves of out of inner conviction, and it stands firmly behind its leadership."[13] Blomberg was clearly signaling Hitler that in contrast to the SA, the army would remain loyal.

On June 30, the "Night of the Long Lives" took place. Hundreds of leading SA officials, including Röhm, were slaughtered by Hitler's henchmen. Meanwhile, the military was not involved. Most generals and admirals felt themselves to be above the conflict—it was an "internal party matter."[14] From the standpoint of the officers, their only complaint was that two army generals (von Seekt and Schleicher) were killed along with members of the SA. On the other hand, the military had reason to be grateful to Hitler. His action had freed the armed forces from the greatest danger it had faced up to that time. The SA was no longer a problem. Its leadership was either dead or in jail. Other members were scattered all over the country. Without a strong leader, there was little chance that the SA would ever play an important role again in German politics. From now on, the military would be the focus of the country's efforts at rearmament. The SS would eventually become a major problem, but for the moment, the danger the SA had presented to the army had been removed.

Hindenburg Dies

On August 2, 1934 Marshall Hindenburg died. Hitler then succeeded in combining the positions of chancellor and president. He was now the highest official in the country. From a legal standpoint, this meant that the country's Nazi Party leader was the military's supreme commander.

At the same time that Hitler became the country's leader, the military, in the form of Marshall Walther von Reichenau, was busy preparing a new oath of allegiance. The old oath (taken during the Wiemar Republic) was to the country's constitution. However, Reichenau and Blomberg believed that if the army swore an oath to Hitler personally, it would bind him closer to the army, show him the military's loyalty, and thereby convince him to stay out of military affairs. In a word, the army hoped that by its taking such a step, Hitler would leave the armed forces alone in the future.

The oath was taken on the same day that Hindenburg died. It read:

> I swear by God this holy oath, that I will render to Adolf Hitler, Leader of the German nation and people, Supreme Commander of the Armed Forces, unconditional obedience, and I am ready as a brave soldier to risk my life at any time for this oath.[15]

The problem, however, was that the distinction between the party and the state had now become blurred. Hitler could theoretically order the army to do anything. There was no control on his actions. Legally, there was no way to defy him. Whatever order he issued (or was issued in his name) had to be obeyed. The idea of an "illegal" order was a thing of the past. If Hitler said it was all right for the army to participate in the killing of Jews, it was not only permitted, it became an obligation.

Hitler was pleased by the army's action, but for the moment he gave no sign that he intended to use it to exert control over the armed forces. As far as the generals were concerned, Hitler and the Nazis continued to talk about the importance of Nazism in the military, and the military took some steps to accommodate him. Meanwhile, the majority of the country's senior officers recognized that they could do little or nothing to stop Hitler even if they had wanted to. He was becoming too powerful. Besides, he was giving them what they wanted. For his part, Hitler realized that despite his position, he had to be careful with the military. As a consequence, he adopted a policy that could have been called "creeping Nazism." He would leave the military alone when it came to major decisions—for awhile. In the meantime, the Nazis would work on new recruits and attempt to gain acceptance of the party and its symbols by members of the armed forces.

Although the German generals thought they were outsmarting Hitler by coming up with an oath to him directly, in time this action would have the opposite impact. As the months passed, the total obedience the soldiers swore to Hitler would be brought up to ensure that they carried out his orders—even when the situation was unethical or hopeless. Regardless of the situation, officers and soldiers would be reminded that they took an oath to obey Hitler—a fact that would also make it increasingly difficult for soldiers to act against Hitler. It mattered little whether or not his actions appeared to violate the constitution. There wasn't any. The only thing that counted was what the Führer wanted.

Infiltrating the Army

The infiltration of the army by the Nazi Party began shortly after Hitler had been named chancellor at the beginning of 1933. For example, on March 12, 1933, the high command issued an order stating that the swastika was to be added to the country's official black, white, and red flag.[16] Then on March 21 a military ceremony was held at the Potsdam Garrison Church on the occasion of the new Reichstag's opening. Hitler was the center of attraction. The military naively believed that by including him, it would be able to publicize the strong role played by the army in the Nazi state. However, Hitler turned the tables on the military. The primary effect was to make it clear to everyone that the armed forces accepted and supported Hitler.

On May 15, Blomberg issued an order stating that all servicemen were to salute uniformed members of the party and its organizations.[17] The order also called on members of the military to salute the colors of these organizations when carried by marching units at public rallies and parades. In addition, Blomberg told soldiers that they must act interested when they were in the presence of party organizations—and he called upon them to join in singing songs at these gatherings. This was a very difficult task for many soldiers, who continued to deeply resent saluting any kind of Nazi organization. After all, in their minds this would put these organizations on an even keel with the military, something that they did not want to accept.

Then in February 1934, Blomberg ordered the Wehrmacht to wear the Wehrmachtsadler (the eagle with a swastika in its talons) on its uniform—symbolizing unity between the party and the military. The military was also ordered to apply the infamous "Aryan" paragraph in appointing military officers and officials. This paragraph was aimed at limiting the number of Jews who could serve in the military.[18] Then on April 21, another order was issued that was aimed at intensifying national socialist training within the armed forces. Soldiers were to demonstrate to the party and the rest of the country that their service was based on "strong loyalty to the national socialist state and the readiness for national socialist training of the soldiers."[19] On April 28, 1934, Blomberg followed up his earlier order by insisting that the national emblem worn on helmets also be worn as a lapel pin, so that when in civilian clothes, military personnel not only would be recognizable but, by wearing this pin, would show their support for the Nazi Party. Then on June 6, 1934, Blomberg ordered the army to initiate closer contacts with party cultural organizations. This was reinforced a few weeks later by another order that told officers to increase contacts with party cultural organizations. By 1934 infiltration of the armed forces by the party was already well under way. This may be another reason why the military accepted Hitler as the supreme commander and did not object to the new oath. The presence of the party and contacts between the military and the armed forces were so advanced that the oath probably seemed normal to most members of the armed forces. Having the party united in Hitler's person and taking an oath to him were normal events in the army.

Hitler worked hard to project an image of a thankful and appreciative leader, who had no intention of interfering in the army's internal affairs and who was prepared to support its special role in the German state. As he stated in a letter to Blomberg in August,

> I wish to express my thanks to you, and through you, to the Wehrmacht, for the oath of loyalty which has been sworn to me. Just as the officers and men of the Wehrmacht have obligated themselves to the new state in my person, so shall I always regard it as my highest duty to intercede for the existence and inviolability of the Wehrmacht, in fulfilment of the testament of the late Field-Marshal, and in accord with my own will to establish the Army firmly as the sole bearer of the arms of the nation.[20]

Officer qualifications also became politicized. In mid-July, candidates for reserve commissions had to convince those in charge of their dedication to the ideals of Nazism. As a secret order issued at that time noted,

> As regards the state, it goes without saying that the Armed Forces accept the National Socialist view. It therefore becomes necessary to convert officers of the reserve to the same way of thinking. In consequence, no one is to be trained for, or commissioned in, the Reserve of Officers unless he sincerely accepts the National Socialist state and stands up for it in public instead of adopting an attitude of indifference or even hostility toward it.[21]

Those entering the ranks of reserve officers would be subject to even more stringent ideological requirements than professional officers. This was a well-calculated action on Hitler's part. As the officer corps expanded, more reserve officers would be called to the colors, thereby increasing the percentage of officers who had been subjected to careful screening by the party. If one phrase embodied Nazi thinking at this time, it was the instruction from 1938 stating that in order to become an officer, an individual must "stand completely on the ground of the national socialist state."[22] The faction in the officer corps that insisted on being above politics—on being apolitical—was becoming increasingly isolated. One could argue that direct party influence was limited by paragraph 26 of the country's military law, which put party membership in abeyance for the period of military service. The same was true of things like promotions, planning, and discipline. All of these decisions were left up to the military, with minimal party influence. This is true, but it is worth noting that despite any formal legal restrictions, more of the country's officers were becoming convinced Nazis.

Nazification of the army continued. In mid-September, the army was ordered to appear at a Nazi Party rally in Nüremberg. The purpose was to show army support for the party—a point pounded home by Goebbels and the party machine. At the same time, Blomberg issued an order permitting soldiers to wear party badges on their uniforms—something that was a sacrilege to many German officers. How could one expect the army to put badges of honor—for courage under fire—on a level with those for party membership or participation in various party activities?

Meanwhile, the party was taking other actions to demonstrate its importance to the military. On September 15, 1935, the Nazi flag was placed on a par with the country's national flag. In March 1936 new flags were issued for the various military units—all of which had the swastika on them. Even more upsetting from the military's standpoint was Blomberg's order of May 26, 1936, which directed that those soldiers who were considered politically unreliable be handed over to the dreaded Gestapo. Another law expanded the Gestapo's authority by decreeing that anyone who was considered recalcitrant or "untrainable" by the military could also be turned over to the Gestapo.

In an effort to change the "hearts and minds" of those in the military and to ensure that the new members stayed convinced Nazis, the party began to push for political indoctrination training within the armed forces. For example, as early as March 1933 the handbook for instruction that had been used up to that time in military schools was withdrawn in favor of "National Political Instruction." The new book was based on the speeches of party officials, as well as on new laws, orders, or speeches by Hitler. Two months later, Blomberg issued a decree, noting that national socialism would become the basis for instruction in all service schools. Regarding the troops, in April 1934 Blomberg issued another order calling for the Discussions of Current Political Questions to be given greater importance. The Ministry of Defense ordered that such sessions be held twice a month in the hope that this would raise the level of commitment to and acceptance of Nazi ideology on the part of members of the armed forces.

From the standpoint of the regular army officer, this was a major change. Up to then, he had been prohibited from dealing with political matters. Now, all of a sudden, company commanders were expected to participate in—if not lead—such discussions. Meanwhile, it was clear that despite the party's efforts to expand its influence in the military, resistance was widespread. For example, when senior party officials gave lectures, "Many big-shot Nazis were laughed at and not taken seriously."[23] Furthermore, German soldiers did not take the regularly scheduled political indoctrination lectures seriously. Blomberg reacted to this situation on April 17, 1935, when he stated, "It has come to my attention that the 'Principles for Instruction in Current Political Matters' are not being given the attention they should receive. I determine what is in them and I authorize their release. They are just as binding as any other official instruction."[24]

By 1936 political indoctrination lectures had become more organized. On January 30, Blomberg ordered that all officer training schools, staff colleges, and the General Staff Academy include a special political indoctrination course in their curriculum. In addition, special training courses were set up in Berlin for those officers who were to become instructors in these educational institutions. The lectures were given by party propagandists. Furthermore, brochures under the title of "Hints for Racial-Hygiene Instruction in the Wehrmacht" appeared. The goal was to produce a soldier who was both militarily competent and ideologically committed. As one source noted,

In the Third Reich, the officer—bound eternally with the German Volk,—must be a convinced national socialist, otherwise he will not be able to fulfill his important task. He must undertake to immerse himself deeply into the national socialist point of view.[25]

The first special lecture series was held in Berlin in 1937. All of the lectures were transcribed and put into book form by the party. The books were distributed throughout the army. Political indoctrination lectures took such priority that by March 1939, the 17th Division reported that although it did not have enough clothing, every soldier had been issued a Nazi instruction book.[26]

The regime soon followed up by organizing weekly instruction for all officers and men in the field. Reading rooms full of Nazi propaganda were set up, and the Propaganda Ministry was authorized to pass out Nazi literature in the military. Each unit was given the money to subscribe to the Nazi newspaper *Völkischer Beobachter*, and books and pamphlets were made available at very low prices.[27]

At the same time that politically approved books and newspapers were made available to the troops, restrictions were placed on what soldiers might read. Material not approved by the party was not permitted in barracks. Those who violated this rule faced disciplinary action. On February 1, 1938, the army set up a censorship office after a number of party hacks complained about some of the things military officers were saying in print. As a consequence, military officers had to submit everything they wrote to this office for clearance prior to publication.

Religion and Chaplains

To officers of the old school, religion was an important attribute of any professional officer. A man might be Lutheran or Roman Catholic; both were acceptable. The key, however, was that he was religious. Indeed, on December 12, 1934, Defense Minister Blomberg had signed an order stating that allegiance to Christianity was a requirement for anyone wishing to become an officer. Blomberg was not alone in this regard. A large number of German officers believed that "the Christian religion was needed as a means for training soldiers and raising combat readiness."[28]

Admiral Räder, the commander of the navy, probably put it best when he noted,

> I very much want you to give the men on my destroyers something that they will be able to take with them on our long voyages. We need all kinds of strength to achieve victory for us, and the coming time. . . . One thing you can be sure of—without the strength that comes from God we will not achieve our goal. That is something we need."[29]

Needless to say, this idea flew in the face of what Hitler and the Nazis sought to produce—a totally atomized military officer, whose first and foremost allegiance was to Hitler and the Nazi Party.

The introduction of conscription in 1935 complicated both the tasks of chaplains and the ideological cohesion of the "old army." Many new recruits were not religious—and had just gone through a course of heavy indoctrination at the hands of Nazi organizations. The same was true of the officer corps. More nonbelievers also entered the officer corps.

From Hitler's standpoint, the situation was not made any easier by the country's two field bishops.[30] Both had serious reservations about some of the actions Hitler was taking. This put Blomberg in a difficult situation. He had to find a way to satisfy party concerns while at the same time keep senior German officers happy. On May 29, 1935, he issued an "Important Political Instruction." In it, Blomberg emphasized that soldiers' attendance at church was voluntary (i.e., they could go to church, but the generals could not force them to attend as in the old days). At the same time, the so-called "Barrack Evening Hours," which were devoted to religious themes, ceased to be obligatory. Meanwhile, a number of Nazi organizations carried out campaigns to convince soldiers to leave their churches and believe only in the Führer. Their God was Adolf Hitler.

As a consequence, chaplains found their activities increasingly restricted. In December, Blomberg again emphasized that chaplains could not use force to compel soldiers to participate in religious services. The distribution of religious material in the military was banned, and only chaplains who were approved by the party could minister to soldiers. As General Dollmann, a senior German military commander who was also a convinced Nazi, observed, "The armed forces, as one of the bearers of the National Socialist state, demand of you as chaplains at all times, a clear and unreserved acknowledgment of the Führer, the State and the People."[31] Furthermore, Dollmann said in a talk to chaplains that they were only permitted to talk to soldiers who voluntarily approached them, and they were strictly forbidden to raise any doubts about the validity of national socialism. From a policy standpoint, the primary impact of Blomberg's actions was to make it clear to the chaplains that while they might be able to count on the support of individual soldiers, they would get little or no assistance from the Wehrmacht in ministering to their flocks. Ideologically, they were not welcome—they placed Christianity and the Christian God between the soldier and his Führer.

In May 1937, Blomberg declared that members of the Confessing Church (the one critical of many Nazi policies) could not serve as chaplains. At the same time, he emphasized that only chaplains in uniform (thereby excluding civilian clerics who often helped out), could minister to soldiers, and he took away the services' control over chaplains and put chaplains under the high command. It was hoped that by recruiting an additional 200 Protestant and 150 Roman Catholic chaplains, it would be possible to replace the civilian chaplains, many of whom belonged to the Confessing Church. These individuals had assisted regular military chaplains because of shortages. This was clearly an overly optimistic plan and, despite efforts by the Wehrmacht, would not be realized by the time war broke out.[32]

It is worth noting that chaplains often came under direct attack from party officials. For example, in 1938 a local Nazi administrator attacked chaplains, claiming that "No chaplain has been in a fox hole. The chaplains are bastards. Christianity and Judaism are the same. He who stands on the Bible has Palestine as a homeland, not Germany."[33] From a policy standpoint, the chaplain's position was always threatened by the willingness of the military to compromise with the party leadership. By the time war broke out, chaplains had been forced to become neutral—the most they could do was give moral comfort to those who came to them. Even then, they had to be very careful to avoid saying anything that might be seen as critical of the Nazi Party or the Führer. Individual soldiers could take part in religious ceremonies, but they were often the subject of ridicule by their compatriots, who accepted the Nazi argument that Christianity was a refuge for weaklings.

The Nazis also vigorously opposed the Freemasons because they, too, belonged to an organization that might try to prevent the Nazi Party's attempt to remove any obstacles between the soldier and his Führer. On May 26, 1934, Blomberg declared that servicemen could not be members of a Masonic Lodge. On October 7, 1935, he went even further. Even former Freemasons were excluded from selection as officers unless they had resigned from their lodge prior to October 1, 1932, and had not taken the Third Degree. Officer candidates had to sign papers declaring that they had never been Freemasons.

Jews

Not surprisingly, one of the blackest pages in Blomberg's many attempts to pacify Hitler and the party was the way that Jewish officers and soldiers were treated. On April 7, 1933, an order was issued that kicked all Jewish officials out of the German government. The military was not immediately affected. A month later, however, military law was amended to make an Aryan background a prerequisite for military service. On December 8, 1933, Blomberg warned the armed forces that although he could not order them not to shop in Jewish stores, "there is a danger of friction if Wehrmacht members continue to make purchases in Jewish shops."[34] Two months later, he went a step further, proclaiming that all Jewish officers would be released from duty. A special exception could be made for those who had served during World War I. A total of seventy officers were released. The exception for those who had served during the war was not to last long, however. All had been dismissed by the end of 1935. In April 1936 the military ruled that a soldier's wife must be of Aryan blood. The next month a decree was issued stating that in selecting soldiers, the strictest racial criteria were to be used.[35]

This is not to suggest that all officers quietly went along with Hitler's anti-Jewish policies in the military. A lot of resistance occurred—primarily in the form of patronizing Jewish shops, as noted previously—an action that often led to fights with Nazi thugs. Furthermore, an effort was made by a number of officers to protect their Jewish colleagues, especially those who had distinguished themselves in combat during the First World War. Nevertheless, by late 1937, the Jewish question had been settled. Jews were not welcome in Hitler's armed forces.

By 1938, the conflict between the military and the party was beginning to come to a head. From Hitler's standpoint, there was too much obfuscation—the soldiers seemed to be going through the motions of accepting Nazism, but their acceptance appeared to be far from enthusiastic. For one thing, the military was so jealous of its command prerogative that it would not let the party have any say in how political indoctrination courses were implemented. From the party's standpoint, there was a big difference between having political indoctrination lectures delivered by military officers—most of whom had little or no interest in the subject—and having them delivered by convinced party officials who preached with the fire of a fundamentalist Christian evangelist. In a nutshell, despite all of its advances, the party was still on the outside looking in.

The Blomberg-Fritsch Affair

The year 1938 was a turning point in party-army relations. Until then, Hitler had succeeded in lulling the generals into believing that although they had made some concessions, they were still in charge. In January, however, Hitler lowered the boom. First, he convinced Blomberg to resign because of his marriage to a woman of questionable moral character. He had brought dishonor to the officer corps. Then, with the help of the SS and the Gestapo, Hitler accused Colonel General Werner von Fritsch, head of the army, of homosexuality—a charge that was clearly bogus. Nevertheless, despite his reputation as a model soldier, Fritsch was placed in a very difficult situation and Hitler relieved him of his duties. He was eventually exonerated, but by that time, Hitler had triumphed with his Anschluß with Austria and no one really cared about Fritsch's situation.

With Blomberg and Fritsch gone, Hitler himself took over command of the armed forces, filling Blomberg's position. He filled Fritsch's position with General Walter von Brauchitsch. In the process, the army—which had been the senior service—was placed on the same level with the navy and Göring's newly created Luftwaffe. No longer did the army enjoy the status it had in the past. In fact, the high command was replaced with the Supreme Command of the Wehrmacht—which included a group of senior military officers who were subservient to Hitler. Instead of professionals working under the directorship of the war minister, whose task was to carry out Hitler's orders while strongly defending the military's opposition, the armed forces were now run "by a body which was nothing more than the military bureau of the Head of State, Supreme Commander, and Commander-in-Chief—Hitler," whose primary purpose was to carry out the Führer's orders and wishes. For this reason, a lot of soldiers referred to the OKW (the German initials for the Supreme Command of the Wehrmacht) as *Oben kein Widerstand* ("No resistance at the top").[36]

The military was now trapped. It had taken a personal oath to Hitler, and the Führer's prestige among the people—as he rolled from one victory to another—made it very difficult to criticize anything he did. The party took advantage of the situation. Von Brauchitsch was asked to "tie the army more closely to the party and to permit greater latitude for the security services within the army."[37] Resisting such pressure would be very hard, especially now that Hitler was the supreme commander of the armed forces. Previously, resistance had meant arguing with a civilian outside of the chain of command. Now it meant taking on the commander in chief. Müller probably stated it best when he observed:

> The leading figures of both the army and the armed forces supreme command now had neither the will nor the ability to represent or apply the concept of an independent role of the army in the state. They no longer aspired to participate in fundamental decision-making or consequently in the power of the state. They confined themselves strictly to professional military matters and to executive functions on the instructions of the head of state and supreme commander of the armed forces—Hitler—who alone issued political directives and assumed responsibility, and who had become their immediate, highest superior.[38]

In short, the balance of power between the army and the party had changed fundamentally. The party could now point out to the army that the military needed it more than the other way around.

The Army Still Has Some Autonomy

In spite of the steps that had been taken by Hitler and the Nazi Party, the military still retained much more autonomy than was the case with the Soviet military and its commissars in the 1920s. Political indoctrination was a fact of life, but these courses were conducted by regular military officers, not by political officers. Furthermore, control lay strictly in the hands of the commander. As one author noted, "Company commanders protected themselves regularly against the construction of an organization of Politruks, which could have undermined their sovereign leadership position."[39] Furthermore, the army had succeeded in maintaining autonomy in the area of investigations. Despite Gestapo involvement in some instances, by and large, it was the army that investigated questions of reliability. Thus, when it came to the issues that really mattered to the generals, the army was still in charge.

> The Army was, therefore, responsible for its own discipline, promotion, training, ideological instruction and direction. No Party organization had the slightest control over the Army whatsoever. Only Hitler, as its Supreme Commander, possessed any authority, but he chose not to use it.[40]

This autonomy was limited by the gradual change that had taken place among both soldiers and officers by 1939. An ever increasing number of them accepted

Nazism and began to revere Hitler; not only had he succeeded in remilitarizing the Rhineland and conducting the Anschluß with Austria, he had also taken over Czechoslovakia. Despite the generals' warnings and concerns, the Führer had once again proved that the country's military leaders were too cautious. As a result, a senior officer would have found it very difficult to stop or undermine Nazi propaganda among his troops. Hitler and his party had certainly turned out to be right at this point in time.

Political Indoctrination

With von Blomberg and Fritsch out of the way, the party began a new assault in the area of political indoctrination. Party leaders began calling upon the military to permit party officials to take charge of ideological training. Not surprisingly, the military responded by noting that such matters came under the province of the armed forces. In a sense, this was the beginning of a push for political officers in the army. The military would be able to stiff-arm the party now, but party radicals, in particular, would not give up the idea.

Beginning in 1938, lectures on national socialism became an integral part of the two-year course of study for officers at the Military Academy. Throughout these lectures, the idea that Hitler, the NSDAP (National Socialist German Worker's Party), and the army were one was hammered into the heads of the students. Then on April 1, 1939, a section for military propaganda was set up in the high command. The purpose was the same as it had been in the past—to show Hitler that the military was loyal without it giving up control over those aspects of military life the generals considered critical for the conduct of war.

Hitler would continue to show his dissatisfaction with the generals, however. On November 23, 1939, he called upon them to show greater ideological conviction. A year later, von Brauchitsch issued instructions dealing with political indoctrination. He made it clear that while the unit commander remained responsible for implementing it, the end goal was a convinced Nazi, even a fanatical one. A soldier must have both qualities—political loyalty and military competence. In short, the purpose of political indoctrination "was to make clear to officers as well as enlisted personnel the purpose of the war."[41] The problem—as it had been in the past—was that a majority of officers thought the instruction was superfluous because they themselves were Nazis. Why issue more political instructions when almost every officer in the army was a member of the party?

The War with Russia and the Commissar Order

The war with Russia was fundamentally different from the battles in the West. War in the West had much more of the gallantry and honor associated with wars in the past. The conflict in the East, however, was a no-holds-barred fight to the finish. The days of gallantry were over. It was an ideological war. From Hitler's standpoint, it was a battle to the death between the forces of good (Germany and Nazism) and evil (Slavs and communism). As he commented on one occasion, "Destroy them or they will make a desert out of Germany."[42] This fanaticism and ideological commitment led Hitler to issue one of his most infamous orders, and one that would inevitably draw the military into the web of barbarism he was weaving. This was the so-called Commissar Order.

Based on this order—which came from the military's commander-in-chief— all Communist Party functionaries and especially all commissars were to be shot immediately upon capture. From Hitler's standpoint, "Political functionaries and commissars as carriers of the Bolshevik idea could not be prisoners of war. After their capture they were to be given over to the SD or where combat conditions did not permit that, they were to be shot by the troops."[43] The military was no longer in a position where it could pay lip service to Hitler's ideological whims while at the same time ignoring them. It was now directly involved in carrying out actions that would later be condemned as war crimes. The same was true with regard to the relationship the military was forced into with the *Einsatzgruppen*, the special units whose primary purpose was to eliminate Jews and other "undesirables" in the conquered territories.

From the army's standpoint, the Commissar Order was a dumb decision. Its primary result was to convince the Russians—and especially the commissars—to fight even harder. "At present, it is so that in any case, the commissar sees a certain end before his eyes, and for that reason a great number fight to the last and also force the Red Army men into the most brutal resistance." The army was so concerned about this decision that officers often "went around Hitler's orders and above all concerned themselves with maintaining discipline by handling matters in a purely military fashion as a way of avoiding political threats."[44] Indeed, there were those in the military who made formal requests for an end to the Commissar Order—but the high command was too weak and too afraid to raise it with Hitler. The generals knew that for him, ideology was more important than the efficiency of military operations.

The high command was well advised to be careful in its dealings with Hitler. On December 19, 1941, Hitler dismissed von Brauchitsch and took personal charge of the army, believing that von Brauchitsch was not doing his job. Hitler doubted the intensity of von Brauchitsch's commitment to Nazism. Relations between Hitler and his generals continued to deteriorate. He could not accept the fact that German troops were losing primarily because of military reasons. In his mind the cause was simple: either they were not fighting hard enough (not fanatical Nazis) or his generals were betraying him. "By 1942, suspicion obsessed Hitler; his behavior was characterized by a complete absence of self-criticism, and an overriding compulsion to see the worst in others, especially in his generals." The situation was so bad that Hitler often bypassed the General Staff and instead issued orders directly to his generals in the field.[45]

Then came the Battle of Stalingrad. On May 8, 1942, the German Army opened a new offensive in the East, with Stalingrad as its goal. The attack was a disaster. As a result of Hitler's stubborn determination to hold onto the city and Herman Göring's inability to supply the beleaguered garrison, the entire 6th Army (300,000 troops with all of their equipment) was forced to surrender on February 2, 1943. Meanwhile, the German Army was in retreat on other fronts—such as North Africa and Sicily.

The question was why? Hitler continued to place constant emphasis on the "will" of his soldiers. He firmly believed that ideological fanaticism would enable an individual to overcome all obstacles. The task was to convince these men of the importance of ideology, to make them willing to lay down their lives for their Führer. On one occasion, he told General Hadler "privately that Nazism would count more than military skill in the future and that he could not expect this from his traditional officers."[46] A method had to found—one that would make German soldiers into fanatics able to overcome any obstacle. In a psychological sense, they were to become "supermen." All of this served to strengthen the argument made by party leaders that regular military training had to be supplemented by more ideological training.

Even prior to the battle for Stalingrad, morale was down throughout the army and especially on the Russian Front. Nothing seemed to be working. "The troops were apathetic and incapable of carrying or servicing their weapons."[47]

This was an issue of concern not only to Hitler, but to line officers as well. Something had to be done to shore up morale among the troops. A way had to be found to make soldiers realize that this was a historic battle between the forces of good and evil. As one officer put it, "We must get to the point where we want to fight, instead of being forced to fight."[48] The problem, however, was that the officers who were closest to the troops and who were supposed to explain the monumental nature of the situation were combat officers—men who could handle themselves well on the battlefield but who knew little or nothing about German history. Hitler and the high command could scream all they wanted about the need for more ideological education, but there was little likelihood that anything would change under existing conditions.

At that point, a proposal was put forth, suggesting that a division morale officer be appointed. This officer was supposed to be both a proven combat officer and an ideologically convinced Nazi. The point, as General Schörner, one of the more ideologically convinced generals, stated in an order he issued, "There is no such thing as a separation between military and spiritual leadership. A soldier of today will win with his weapons and ideology."[49] The concern on the part of the regular military was that the morale officer would be a stalking-horse for the introduction of a commissar system as existed in the Red Army. Commanders were worried that they would have to share authority with political hacks. No line officer wanted to see such a situation. While the military went ahead with the proposal—in the form of morale officers, not political commissars—it took quite awhile to get this idea implemented. Line officers did not resist openly, but they also did not make the introduction of such officers a priority. In the end, morale officers do not appear to have taken up their duties until the middle of 1943. Commanders were responsible for seeing that ideological training was carried out in their units.

It is an irony in the evolution of the National Socialist Leadership Officer that even though Hitler ordered the Soviet commissars killed upon capture, he had great admiration for them and what they had accomplished. Indeed, one gets the impression that Hitler would have loved to have had political commissars in the Wehrmacht, but he feared that they would bring irreparable harm to military efficiency.

Seen from Hitler's perspective, the political commissars had done a lot to advance the Soviet cause. He realized that their knowledge of technology was limited, but, as Goebbels noted, "in him glowed the fire of a revolutionary idea. And with holy dedication he carried out his new task."[50] The problem from the standpoint of the Nazis was how to come up with something like the commissar—without damaging military efficiency. Chaplains helped in Cromwell's army, but given the Nazi prejudice against religion and the fear that it would get in the way of devotion to Hitler, it was doubtful that chaplains would be able to play a significant role in this area—even if they wanted to.

Chaplains and Religion during the War

One thing was clear—chaplains could not fulfill the motivating role the Nazis were bent on creating. At the same time, the Nazis realized that, like it or not, chaplains played an important role in maintaining morale in the military. The Nazis could not simply remove them and religion from the armed forces. Millions of Germans still held to their belief in God and were active members of the various churches. Trying to purge them of their ideas and beliefs would be counterproductive and would certainly undermine military efficiency. And there were "patriotic" chaplains who supported Hitler. In 1939, for example, the Bavarian Bishop Meiser demanded that chaplains in the army preach that "our belief is the victory, that has overcome the world." In the same year, Protestant chaplains received from their churches a pamphlet that stated as follows:

> Bless the Führer of our people and all of his helpers and advisers. Be with the entire German Wehrmacht, bless German weapons and give them victory. . . . In your hands we place our military leaders on land, on water, and in the air: give our weapons victory.[51]

This is why in 1939 the high command issued an order, noting that chaplains would continue to play an important role in the German Army. To quote it,

> All military experience has shown that the spiritual power is an army's strongest weapon. It takes its power in the first instance from a strong belief. The chaplain is for that reason an important means for strengthening the combat capability of our army.

Chaplains are a critical part of the Wehrmacht, and will more effectively meet the demands and the maintenance of internal combat effectiveness to the degree that they are able to satisfy the spiritual needs of the German soldier in the field.[52]

In fact, chaplains were most important to the army. As a German chaplain put it in a somewhat humorous fashion, "The navy tolerated chaplains for reasons of tradition, the army needed them to strengthen the morale of the troops, but the air force was so close to heaven that they felt they could abstain from chaplains."[53] The bottom line, however, was that the services could not do without chaplains when it came to morale. Chaplain Eich provided an example of the kind of morale services that chaplains provided. In one instance, he was called to a naval base because a sailor had committed suicide. The commander ordered Chaplain Eich to do something about the situation that led to the sailor's suicide—"every sailor is needed." He quickly discovered the problem and fixed it.[54] Chaplain Eich commented several times in his book on the role he played at executions. On more than one occasion, Eich was called upon to spend time with a condemned soldier or sailor and then to accompany him to his execution—clearly, a task that a leadership officer was not about to carry out.[55]

Still, the Nazi Party remained suspicious of chaplains—even those who declared their loyalty to the regime—and did little more than tolerate them. Those who were believed to be disloyal or even potentially unreliable were thrown out of the military. For example, on May 31, 1941, all Jesuits (who were considered a special threat) were released from their duties as chaplains. A year later, an instruction was issued, making it clear that chaplains were to support national socialism and that they could only visit wounded soldiers if the soldiers specifically requested such a visit.[56] To make matters worse, chaplains could only utilize religious material approved by the high command. Then, as if to underline the point, an order issued on September 17, 1942 emphasized that participation in all religious services was voluntary.[57] In spite of this order, German chaplains found that certain commanders would play games with them and their services. One chaplain commented, for example, that "when on Sundays I appeared with my 'Church in a Suitcase' the men were suddenly ordered to take part in some important exercise."[58]

In addition to helping raise morale, chaplains played a role in another area, somewhat similar to that in Cromwell's army, by holding religious services on the eve of battle. While most officers were not religious, they understood that such services could play an important role in "strengthening the inner strength and steadfastness of soldiers." Indeed, refusal to conduct these services and "a failure to use chaplains in the field there would have done unacceptable harm."[59] The high command considered the chaplains' primary task to be raising combat readiness. As the high command noted on February 9, 1942,

The main task of the field chaplain is and remains . . . strengthening the combat capability of the troops. . . . The field chaplain must always be and remain a means to an end and must not become an end in himself. The field chaplains must under no circumstances represent the one sided self-interest of the church, but to the contrary, he must help the German soldier find the internal strength, which the man at the front needs to carry out his difficult task. Like every German, the chaplain must devote his entire effort toward the great goal of winning this war.[60]

The patriotic chaplains (in contrast to those from the Confessing Church) tended to agree with the party's call that it was their duty to convince the German fighting man to give up his life for his country. As one clergyman from the Patriotic Church put it, "Such soldiers can fight in comfort and undaunted because God is with them. Such soldiers can return home in peace because God's arms await them."[61]

The Nazis would continue to be suspicious of chaplains to the very end. One task that would be assigned to the new National Socialist Leadership Officers would be to watch over the actions of the chaplains to ensure that they did nothing to undermine the war effort. But this effort was not always successful. Chaplain Eich provided an interesting comment on the relationship between chaplains and NSFOs. When the word came down that his unit had to have an NSFO, the commander determined "it almost impossible to find an appropriate officer for this office." Eventually, he located an officer who was an outspoken opponent of the regime, a man who called himself a PG, or enemy of the party (*Parteigegner*). When Eich asked him why he selected such a man, the admiral said, "Because no one has anything to fear from him." According to Eich, "This officer was of great assistance. He did not turn anyone in. I am still thankful to him that I was able to carry out my work without fear."[62]

Regardless of how much some Nazis might not have wanted the chaplains around, these clerics were indispensable to the war effort. As one author noted,

> By reclaiming God for the German side, by creating a willingness to resist, by giving the promise of a reward in heaven, and a willingness to sacrifice one's self, a tremendous contribution was made to maintaining troop morale. The high evaluation that the military paid to their role both during the war, and after it confirms this conclusion.[63]

The bottom line was that chaplains were crucial in keeping morale up, especially in the army.

The Army and Jews

The army's association with Hitler's policy of eliminating Jews would continue to haunt the German Army, both during and after the war. Regardless of the fact that many officers had little time for Hitler's policies vis-à-vis the Jews, in reality the armed forces were closely associated with the SD and the so-called *Einsatzgruppen* throughout the war. Because of their personal oath to the Führer and the Commissar Order, soldiers were in no position to refuse to cooperate with the *Einsatzgruppen* as they carried out their morally repugnant work. In fact, especially on the Russian Front, the army and the SS worked hand in glove in an effort to annihilate as many Jews and other undesirables as possible. And special efforts were taken to make sure that this process worked smoothly.

During the invasion and occupation of Poland, a number of officers had protested against the use of the *Einsatzgruppen*. Not only did military officers find the actions of such units morally unacceptable, they also opposed any involvement by the armed forces in such activities. With this in mind, detailed arrangements were made at the highest level to prevent any opposition on the part of the army when Berlin invaded the Soviet Union. In short, by 1942 the military was up to its armpits when it came to the elimination of the Jews and other undesirables. "Not content with shifting responsibility on the murder squads, the army co-operated actively at all levels first in the shooting of communist commissars and Jews and later in the exploitation of Russian POWs as slave labor in Germany"[64]

Führerpersönlichkeit

While it may sound a bit silly to American and British ears, what the Nazis were trying to do was to create an army made up of little Führers—soldiers who were so indoctrinated that they would think and act like Hitler and have the Führer's personality (*Führerpersönlichkeit*). As the ramifications of a losing war on the Eastern Front began to sink in, it became increasingly obvious that something had to be done. If Germany were to win the war, the "will" of the German soldier would have to be strengthened. More Germans had to be made into little Hitlers. The question, however, was how to do it? First, political indoctrination courses were strengthened. An order issued on May 22, 1942 stated that more effort should go into making soldiers aware of the awesome responsibility they carried—for the future of Germany and the Aryan race. The order emphasized that only if such courses were taught by convinced Nazis would it be possible "to release the combat energy of the soldiers."[65] But there was general agreement that would not be enough. Maybe the approach taken by the Soviets in introducing political officers had something to it.

Political Officers?

The idea of a political officer was not a new one in Germany. For example, during the First World War, especially during the later stages as the situation became increasingly difficult for the Reichswehr, "the high command assigned certain officers to talk to the troops in order to refurbish units even in the remotest places and to counter the destructive tendencies." These officers were called "Officers for Fatherland Instruction."[66] Their task was to improve morale among soldiers at the front.

Then in November 1919, the navy mutinied in Kiel and there was a great danger that the revolt would spread to other parts of the German military. In order to keep soldiers in line and free from the influence of "revolutionary ideas," soldiers were made into *Bildungsoffiziere* (Education or Training Officers), whose task was to talk to their comrades in an effort to maintain discipline. Adolf Hitler was one of these officers. In fact, this was the primary reason why Hitler avoided being discharged in 1919 as the military was reduced in the aftermath of the war. His work as a *Bildungsoffiziere* made him indispensable and kept him in the army until March 1920. He worked primarily for the propaganda department of the government. In May 1919, after the communist republic had been crushed in Bavaria, a lot of work remained to be done when it came to indoctrinating soldiers. Ian Kershaw described the formidable tasks facing individuals such as Hitler as follows:

> The education of the troops in a "correct" anti-Bolshevik, nationalist fashion was rapidly regarded as a priority, and "speaker courses" were devised in order to train "suitable personalities from the troops" who would remain for some time in the army and function as propaganda agents (*Propagandaleute*) with qualities of persuasion capable of negating subversive ideas.[67]

Hitler was in his element! He was doing something he was very good at, and he claimed that he brought "many hundreds, probably thousands" of soldiers "back to the Fatherland" as a result of his lectures.[68]

Given his personal background as what amounted to a political officer, it was not surprising that he supported the idea of morale or political officers in the Wehrmacht when the issue was raised. For its part, the army tried everything to ensure that these individuals would not intrude on the commander's ultimate responsibility for his unit. In 1942, divisional morale (*Betreung*) officers had been established. It was their task to oversee not only morale, but also propaganda in the unit. The morale officer worked with unit commanders and supplied them with such material, while at the same time stressing to both NCOs and officers the importance of presenting a "National Socialist image and example for their men to follow."[69] By June, the high command had ordered all units to create Officers for Military Spiritual Leadership (*Offiziere für Wehrgeistige Führung*) down to regimental and battalion levels. A month later, Field Marshal Wilhelm Keitel ordered commanders not to use chaplains as Spiritual Leadership Officers. The reason was simple. Chaplains were an anathema to Nazis. How could they be expected to impart the kind of fanaticism that was needed to win the war? Yet commanders were only too happy to use chaplains for this type of work. Why use a line officer for such a task? Besides, from a commander's standpoint, this was chaplains' work, one of their primary functions. And it was not only chaplains whom commanders tried to use in this capacity. They also tended to assign their least useful officers to this function. Recognizing this problem, Keitel wrote to commanders on July 15, insisting that they improve the quality of such officers. Only those who possessed "spiritual alertness, self-reliance, good judgement, an appreciation for political activity, and a strong belief in questions of Nazi ideology," should be selected for this job.[70]

Creating a National Socialist Leadership Officer (NSFO)

Given its druthers, the army would have preferred to keep the morale or education officer arrangement. It permitted the military total control over the education officers' activities. Political propaganda might come from outside the armed forces, but the commander on the scene decided what would be said to the troops and when such lectures would take place. After all, the commanders had guidebooks prepared by the OKW and, in the eyes of many officers, this was all that was required for them to provide ideological training. Besides, as the army never stopped pointing out, SS troops did not have officers who were specially designated to carry out ideological indoctrination. That was the task of each and every SS officer.

The next step in the creation of a Nazi political officer came on May 14, 1943. The high command ordered the military to appoint officers whose only task would be military-ideological training. Up to that point, an officer carried out this job in addition to his regular duties (which generally meant that it was ignored). In addition to working in these areas on a full-time basis, officers were expected to visit subunits and report on morale monthly. In defining the characteristics of this officer, the May 14 order stated that such an individual should "be an officer who has proven himself at the front, be an active National Socialist, be a person with a lively personality, be able to unite his actions and deeds and able to transmit them to others, be excited about his work." The order also noted that the rank of the officer was unimportant.[71] While members of the military may not have realized it at the time, the foundation for the NSFO had been laid.

In October 1943, Hitler gave a speech to commanders and morale officers, in which he reiterated a theme that had long been a dogma among Nazi Party members. He noted that given the circumstances on the Eastern Front in particular, it would no longer be possible for an officer to be only a good specialist in military matters; he must also be politically active and able to lead his troops with ideological enthusiasm. Hitler made it clear, however, that he was not talking about creating political commissars based on the Soviet system. Nonetheless, it was up to every officer to become a "spiritual" leader.[72]

On November 28, 1943, the morale officer became the "National Socialist Leadership Officer." His position in the military hierarchy was that of a deputy to the unit commander. Just as the Ia officer was responsible for tactical matters, the NSFO was responsible to the commander for propaganda and ideological leadership. His position was not much different from that of a Soviet political officer. He had no control over the actions of the commander (the way a political commissar did), but instead, he was clearly subordinate to the commander. His job partly dealt with inculcating the troops with Nazi ideology. Like his Soviet counterpart (at this point in time), it was also his job to talk with the troops and answer their questions. Hitler did not want an aloof propagandist. Rather, he wanted someone who knew the troops, an officer who could gain their respect and who could interact with them in such a manner that he could "politically activate the troops and instill in them a Nazi will for victory."[73] Morale building was also an important part of this officer's job.

On December 22, 1943, Hitler issued an order that formalized many of the points noted previously. Then on December 30, a memorandum was prepared by party leaders that sharply attacked the situation as it existed at that time. This memo maintained that it would be a disaster if the existing morale officers were turned into NSFOs. The memo argued further that this was exactly what the military would try to do. Time had shown that the current crop of morale officers did not have the right "qualities." In the past, the military had relied on individuals who were not suited for other military duties to be morale officers—for example, teachers. In some instances, even officers who had studied Christian theology were utilized! NSFOs should be regular officers—those who were not only convinced but fanatical Nazis. Only fanatical, well-trained officers would be able to turn soldiers at the front—who were now being defeated on an almost regular basis—into fanatics.[74] The only hope was to make the army a model of the Waffen SS.

In January 1944, Hitler held a conference with several of the key players in the creation of NSFOs. As a result of this conference, the roles and functions of these officers were further clarified. First, NSFOs were to conduct their work via military channels—not through party channels. NSFO positions would be created down to the division level. Some individuals from the former *Wehrgeistige Führungs* officers would be utilized, but only if they were neither intellectuals nor undesirables. Given the shortage of officers that existed in the German Army during the war, regimental and battalion officers would have to perform this function in addition to their other duties. Finally, Hitler emphasized that the commander would retain final authority. One can surmise that, based on his comments, Hitler had certain misgivings with regard to how vigorously this policy would be implemented because he noted that his own experience as a political officer demonstrated that commanders would undermine such work whenever possible.[75]

Hitler's comments were given the force of law in the military in a set of Provisional Regulations for NSFOs published the next month. This document stressed that it was the commander's responsibility to be the "bearer of national socialist leadership in the Wehrmacht." The German Army continued to insist that the commander in the field, rather than a commissar, was the primary representative. The NSFO was only there to assist him whenever necessary. Then in July, Colonel General Heinz Guderian reinforced this idea in an order in which he maintained that every staff officer should act like an NSFO.[76]

In an effort to get things started, General Hermann Reinecke was put in charge of NSFOs. The problem, however, was that General Reinecke had never served at the front. Instead, almost all of his experience came from sitting behind a desk. Not surprisingly, line officers often had little respect for him and his efforts to motivate the troops.

Most army field commanders worked overtime to sabotage the program. They continued to strongly resist pressure to send their best officers back to Germany for the course that was set up to train and evaluate NSFO candidates. When they did send officers, they continued to send individuals who had been theologians or teachers in civilian life—even though they knew that such officers would be unacceptable. If nothing else, this kept their best officers in their units at a point in time when they were fighting for their lives. The high command soon discovered that if it did not transform the morale officers into NSFOs, there would not be enough officers to fill the open slots.

Training NSFOs

The school for NSFOs was set up at Krössinsee, a former party school located near Stettin in Pomerania. The first course began on March 8, 1944 and included 227 students. There were 208 officers from all branches and 19 civilian workers. For the first time, it was possible to get officers from all three services together at once to hear ideological lectures. The course lasted 12 days. Of the 227 students, 80 percent were considered suitable for service as NSFOs. The majority had already been morale officers.[77]

The second course was held from March 17 to 27 and upset Berlin. A number of participants had announced ahead of time that they did not want to be NSFOs (even before the class started). Furthermore, many of those who showed up were again teachers and intellectuals, just the kind of individual the Nazis did not want. Additional classes were held, but the same complaints were heard from students— the speakers had not been in combat. It was easy for them to talk about the importance of giving one's life for the Führer and the Fatherland, but what did they know about the privations of life at the front? The school administration also constantly complained about the quality of the students. It was clear to the school's leaders that they were being sent the weakest and the least useful officers—exactly the opposite of what they believed was necessary.

To get an idea of how confrontational the situation was, consider the class that was held from November 14 to 28. One hundred eighty-two students from the War Academy in Hirschberg attended part of the course. Their commandant accompanied them and they left after the first eight days as planned. From the beginning, these future general staff officers were a problem. They were visibly aloof and showed little willingness to conform to the demands of the course. They refused to take part in mandatory group work sessions. One general staff colonel who was a part of this group called the course "grotesque." Furthermore, most of these officers let it be known on the first day that they already had mastered this "stuff" and were only attending for informational purposes.[78]

During the time NSFO classes existed (they were closed on February 17, 1945, and moved near the Olympic Stadium in Berlin), 2,435 students attended them. Of that number, 1,766, or 72 percent, were accepted as NSFOs. Unfortunately, demand was far greater than supply. The majority came from the army, with the navy and air force a distant second and third. There were a total of 469 regular officers and 1,601 reserve officers. Only 972 lacked any religious affiliation, indicating how difficult the problem was: If many of the NSFOs were religious, and these were the party's stalwarts, what chance did the party have of wiping out religion?[79]

Within the ranks of the regular army, NSFOs were looked upon as a temporary measure. Given the current situation, the decision to create them made some sense. Once the war was over, however, the military planned to get rid of them. The NSFO's purpose was not ideological in the military's eyes. Rather, these officers were there to raise the spirit of the troops so that they would fight better.

The attempt on Hitler's life served to strengthen the hand of the NSFOs. Many Nazis saw this as further proof that the officer corps was disloyal and that more ideological indoctrination was needed. However, the most immediate impact of the assassination attempt on Hitler was the introduction of the Hitler greeting throughout the Wehrmacht (the outstretched arm, rather than the more traditional military salute). Hitler also played with the idea of permitting NSFOs to evaluate politically all officers, including their superiors. In the end, this order was not implemented.

As the war worsened, the task of NSFOs moved further away from education and indoctrination and more toward preparing soldiers for battle. As the order of August 3, 1944 put it, "It is no longer the task of the NSFO to educate the soldier"; now the NSFOs' task was "to devote themselves to the most extreme activation and fanatizization of the troops." Or as it was stated in another context, "The German officer must be an absolute believer. He who does not fight today with his belief, he fights with only half power."[80]

The closest the NSFOs came to playing the role of a commissar was on March 13, 1945, shortly before the Nazi regime collapsed. The new order, issued on that date, stated that in addition to his task of whipping up fanaticism and political activity on the part of the troops, the NSFO was to play an important role as an informant. If soldiers believed that someone was unreliable, they were to report it to the NSFO, who in turn had the responsibility of informing the security organs. As a result, if the NSFO really wanted to get his commander in trouble, he had the opportunity of claiming that the commander was unreliable—a very risky game, however, because if the NSFO was wrong, he could find himself on trial.

During the last months of the war, German troops were consistently on the defensive. The Third Reich was crumbling. Rather than functioning as the commander's deputy in matters of ideology, the NSFO was forced to live in a world of confusion, if not total chaos. Often, the orders were contradictory and made little

sense. Meanwhile, the Nazi regime, including the army and the party, was collapsing. In such a situation, the NSFO was never able to play the role Hitler marked
out for him. Indeed, by the time Hitler committed suicide, most NSFOs were already in American, Russian, British, or French prisoner of war camps. It is however, worth noting that from December 1943 onward, when the NSFO experiment
was set up, some 5,000 Wehrmacht officers served as NSFOs.[81]

Evaluating NSFOs

But how effective were they? Did they actually succeed in increasing the willingness of German soldiers to lay down their lives for their Führer? To be honest,
the NSFOs faced a difficult task. If they did nothing but repeat party ideology,
they would quickly lose credibility. One did not have to be a genius to know that
the tide of war had turned against the Germans. According to one author, if the
NSFO hoped to be effective, it was critical for him (1) to avoid official statements
that no one believed, and (2) to be a "good guy" (*prima Kerl*).[82] There was no
room for "100 percenters" on the battlefield. Indoctrination lectures were often
counterproductive. To be effective, the NSFO had to spend a lot of time in personal discussions—and there were by no means enough NSFOs available to carry
on such discussions in small groups.

Major Walter Fellgiebel, who was chairman of the West German Association of
the Wearers of the Knight's Cross (one of the highest decorations in the Wehrmacht), responded to a question concerning the effectiveness of the NSFOs after
the war by stating that the program was unsuccessful:

> He said that in most cases the NSFOs were hated very much by the officers and men
> if they took their duties at all seriously. He felt that the hour was simply too late for
> the NSFOs to have any effect on the war; the mass of soldiers had had enough and no
> longer believed in final victory. "Anyway, it was also clear that the war could only be
> ended if the Nazi regime disappeared. On these grounds alone, the NSFOs could not
> change the course of the war as Hitler hoped."[83]

In correspondence with Quinnett, Field Marshal Erich von Manstein made the
same type of comments. "It was too late. One cannot conduct any war at the greatest intensity for six long years without the armed forces and people suffering more
and more damage. This fact was already discovered in the First World War. The
thoughts of Hitler and the party to change this situation through the NSFO were
naive."[84]

It is also important to understand that the NSFOs were not taken very seriously
by the troops. NSFOs knew very well that their ability to act—and to survive—
would be more influenced by how much they were able to help the troops with
real, personal problems. This is why NSFOs spent more and more time helping
with the welfare of soldiers. Many soldiers' families were facing serious problems
at home as a result of Allied bombing, and since the activity of chaplains was so

restricted and line officers were primarily concerned with fighting battles, the burden of helping soldiers with personal problems often fell on NSFOs. The good NSFO worked hard to cooperate with the soldiers in his unit. And as far as raising the fighting spirit was concerned, NSFOs had some impact on the German soldier's will to resist the enemy. NSFOs did not, however, achieve the kind of miracles that Hitler had expected.

Conclusion

The German case is instructive for a number of reasons. To begin with, it shows that if the political goals being pursued by the political leadership are important enough, and if time is also an issue of importance, even a highly ideological regime like Nazi Germany is prepared to live with a semi-autonomous military. It was an act of ideological expediency. Conquering Europe was more important to Hitler than purifying the German officer corps.

Besides, Hitler was aided by both the naiveté of the old line officer corps and the fact that a large number of German officers supported him. After all, even if they did not like the way he did things, he was the key that would unlock the door to remilitarizing the country. He was expanding the military—almost overnight—and that meant more promotions and a greater chance for Germany and the army to overcome the shame of Versailles.

By underestimating Hitler—and grossly overestimating their own abilities—Germany's senior officers also made it clear to the Führer that they could be manipulated. They thought they could handle him, when in fact the situation was reversed. Given that Hitler was in the driver's seat from the time that Hindenburg died, why worry about purifying the military? The army had taken a personal oath to him, and he knew that by this act they had irreversibly bound themselves to him. For all their bluster and aristocratic airs, it was the former house painter from Austria who was in charge. He knew it, and too late they became aware of it.

In the meantime, Hitler and his cronies focused on gradually—but irreversibly—increasing their influence within the military. And they succeeded. Like the plague, Nazi influence was everywhere. The party systematically politicized the army, even if not at a very fast pace. Today, new flags, tomorrow enforced ideological instruction and morale officers. The bottom line was that gradually, senior military officers were being isolated. Besides, Hitler was also in the process of building up a parallel military force, the Waffen SS. It is a matter of speculation, but one has the impression that if Germany had won the war, the Waffen SS would have gradually devoured the Wehrmacht. In this way, Hitler would have obtained his highly politicized armed forces.

Just as Hitler was pragmatic in his approach to dealing with senior officers in the Wehrmacht, he was equally flexible in handling chaplains. He had no use for them or their dogma—which got in the way of total devotion to his person. Nevertheless, Hitler recognized that they served a useful purpose when it came to keeping the average soldier on the battlefield. In the end, who cared if the soldier died fighting for the Christian God or Hitler—as long as he was a fanatical soldier? When it came to building morale and, to a lesser extent, motivation, chaplains played an important role. In this sense, the chaplains and the morale officers (including the NSFOs) shared the tasks of maintaining morale, providing motivation, and ensuring political socialization.

Given his own experience as a one-time political officer, it is not surprising that Hitler listened to those who suggested the creation of NSFOs as a means for restoring morale and fighting spirit. The problem, however, was that by the time these officers had been established and given what they needed to do their job, it was too late. The military never really took this idea seriously, the positions were never completely filled, and as NSFOs took to the field, they were faced with an impossible task of convincing soldiers to fight and die for a regime that everyone knew was on its last legs. In this sense, Hitler's pragmatic streak again came through. He could have given the NSFOs the kind of authority enjoyed by Soviet political commissars. He recognized, however, that this would not only be strongly resisted by the military, it would also undermine military efficiency—something that he could not afford.

The relationship between Hitler and the Wehrmacht was a marriage of convenience. Hitler would have preferred to get rid of the conservative generals who had doubts about him and his way of conducting war. On the other hand, Hitler decided he could live with an "impure" officer corps—first, because he needed them to achieve his goals in a limited amount of time, and second, because they were not overly hostile; they merely wanted autonomy to conduct their own internal affairs. Besides, Hitler knew that as the army was expanded and the party intensified its policy of "creeping Nazism," it would only be a matter of time before the officer corps became Nazi, both in spirit and in action.

The creation of the NSFOs was only a part of this larger policy. If the war had gone better, NSFOs might never have appeared on the horizon. They were a stopgap measure and not a very good one at that.

Notes

1. As far as I can determine, the only article/book-length study available in English on this topic is in the form of a dissertation. See Robert Lee Quinnett, "Hitler's Political Officers: The National Socialist Leadership Officers," University of Oklahoma, 1973.

2. A similar situation existed in the post–World War II communist Polish Army. In that instance, chaplains were permitted a very limited role within the armed forces. At the same time, the Polish armed forces had political officers. See, for example, Johann Black, *Militärseelsorge in Polen* (Stuttgart: Seewald Verlag, 1981), and Stanislaw Chmielswski, *Wspomenienia Kapelana* (Suwatki: Wydawnictwo Hatza, 1992), for a fuller explanation of the role played by Polish chaplains in World War II, as well as in the Polish People's Army.

3. Quinnett, "Hitler's Political Officers," 5.

4. Manfred Messerschmidt, *Die Wehrmacht im NS-Staat; Zeit der Indoktrination* (Hamburg: R. v. Decker's Verlag, 1969), 19.

5. Robert J. O'Neill, *The German Army and the Nazi Party, 1933–1939* (London: Cassell, 1966), 15.

6. Klaus-Jürgen Müller, *The Army, Politics and Society in Germany, 1933–45* (Manchester: Manchester University Press, 1987), 33–34.

7. As quoted in Matthew Cooper, *The German Army, 1933–1945* (Lanham, Md.: Scarborough House, 1978), 28.

8. O'Neill, *The German Army and the Nazi Party*, 62.

9. Michael Salewski, "Die bewaffnete Macht im Dritten Reich 1933–1939," in *Handbuch zur deutschen Militärgeschichte* (München: Vernard & Braefe Verlag für Wehrwesen, 1978), 66.

10. O'Neill, *The German Army and the Nazi Party*, 34.

11. O'Neill, *The German Army and the Nazi Party*, 38.

12. O'Neill, *The German Army and the Nazi Party*, 40.

13. Messerschmidt, *Die Wehrmacht im NS-Staat: Zeit der Indoktrination*, 29.

14. Salweski, "Die bewaffnete Macht im Dritten Reich," 78.

15. As quoted in O'Neill, *The German Army and the Nazi Party*, 55.

16. Messerschmidt, *Die Wehrmacht in NS-Staat*, 31.

17. In contrast to other countries, in Nazi Germany everyone seemed to be wearing a uniform. It was for this reason that some members of the Nazi Party—in their fancy uniforms—would insist that members of the military show them the appropriate courtesy by saluting them. Needless to say, from the military's standpoint, such an action was absurd. Why should a veteran combat soldier salute a clerk from the Foreign Ministry? The latter had never seen combat and in all probability never would. A salute was an act of respect from one soldier to another.

18. For a copy of this order, see O'Neill, *The German Army and the Nazi Party*, 38.

19. Messerschmidt, *Die Wehrmacht im NS-Staat*, 34.

20. As quoted in O'Neill, *The German Army and the Nazi Party*, 58.

21. As quoted in Cooper, *The German Army, 1933–1945*, 33.

22. As quoted in Messerschmidt, *Die Wehrmacht im NS-Staat*, 146.

23. Salewski, "Die bewaffnete Macht im Dritten Reich, 1933–1939."

24. As quoted in Cooper, *The German Army*, 45.

25. As quoted in Messerschmidt, *Die Wehrmacht im NS-Staat*, 77, 163.

26. O'Neill, *The German Army and the Nazi Party*, 72.

27. Quinnett, "Hitler's Political Officers", 21.

28. Jens Müller-Kent, *Militärseelsorge im Spannungsfeld zwischen kirchlichem Auftrag und militärischer Einbindung* (Hamburg: Steinmann and Steinmann, 1990), 19, 21.

29. As cited in Müller-Kent, *Militärseelsorge im Spannungsfeld zwischen kirchlichem Auftrag und militärischer Einbindung*, 23.

30. A field bishop is the senior chaplain in the army. In the German Army there were two: the senior Roman Catholic chaplain and his Protestant counterpart.

31. As quoted in Cooper, *The German Army*, 35.

32. Messerschmidt, *Die Wehrmacht im NS-Staat*, 190.

33. Messerschmidt, *Die Wehrmacht im NS-Staat*, 192.

34. As quoted in O'Neill, *The German Army and the Nazi Party*, 36.

35. Cooper, *The German Army*, 36, 75.

36. Cooper, *The German Army,* 86, 87.

37. Messerschmidt, *Die Wehrmacht im NS-Staat,* 213.

38. Müller, *The Army, Politics and Society in Germany 1933–45,* 37.

39. Volker R. Berghahn, "NSDAP und 'Geistige Führung' der Wehrmacht 1939–1945," *Vierteljahrshefte für Zeitgeschichte* 1 (January 1969): 19.

40. Cooper, *The German Army,* 45.

41. Arne W. G. Zoepf, *Wehrmacht zwischen Tradition und Ideologie* (Frankfurt am Main: Peter Lang, 1988), 40.

42. As cited in Berghahn, "NSDAP und 'Geistige Führung' der Wehrmacht 1939–1940," 33.

43. Messerschmidt, *Die Wehrmacht im NS-Staat,* 400.

44. Messerschmidt, *Die Wehrmacht im NS-Staat,* 400, 406.

45. Cooper, *The German Army,* 441, 442.

46. Quinnett, "Hitler's Political Officers," 36.

47. Berghahn, "NSDAP und 'Geistige Führung' der Wehrmacht, 1939–1943," 33.

48. As cited in Berghahn, "NSDAP und 'Geistige Führung' der Wehrmacht, 1939–1943," 42.

49. "Zur Geschichte des Nationalsozialistischen Führungsoffiziers" (NSFO), *Vierteljahrshefte für Zeitgeschichte* 1 (January 1961): 78.

50. As cited in Berghahn, "NSDAP und "Geistige Führung" der Wehrmacht, 1939–1945," 48.

51. Müller-Kent, *Militärseelsorge im Spannungsfeld zwischen kirchlichem Auftrag und militärischer Eingindung,* 28.

52. As cited in Müller-Kent, *Militärseelsorge im Spannungsfeld zwischen kirchlichem Auftrag und militärsicher Eingindung,* 25.

53. Franz Maria Eich, *Auf verlorenem Posten?* (Aschaffenburg: Paul Pattloch-Verlag, 1979), 17.

54. Eich, *Auf verlorenem Posten?* 80.

55. Eich, *Auf verlorenem Posten?* 141–142.

56. The instruction is contained in Dieter Beese, *Seelsorger in Uniform, Evangelische Militärseelsorge im Zweiten Weltkrieg* (Hannover: Lutherisches Verlagshaus, 1995), 88–89.

57. Messerschmidt, *Die Wehrmacht im NS-Staat,* 283.

58. Eich, *Auf verlorenem Posten?* 17. The "Church in a Suitcase" referred to the fact that German chaplains (like almost all other chaplains) carried their important religious instruments in a suitcase.

59. Messerschmidt, *Die Wehrmacht im NS-Staat,* 288.

60. As cited in Messerschmidt, *Die Wehrmacht im NS-Staat,* 301.

61. As quoted in Müller-Kent, *Militärseelsorge im Spannungsfeld zwischen kirchlichem Auftgrag und militärischer Eingindung,* 31.

62. Eich, *Auf verlorenem Posten?* 168.

63. Eich, *Auf verlorenem Posten?* 33.

64. William Carr, "Introduction," in Müller, *The Army, Politics and Society in Germany 1933–1945,* 7.

65. Messerschmidt, *Die Wehrmacht im NS-Staat,* 307.

66. Quinnett, "Hitler's Political Officers," 252.

67. Ian Kershaw, *Hitler, 1989–1936, Hubris* (New York: W. W. Norton, 1998), 122.

68. Adolf Hitler, *Mein Kampf* (Boston: Houghton Mifflin, 1939), 289.

69. Quinnett, "Hitler's Political Officers," 47.

70. Quinnett, "Hitler's Political Officers," 54.

71. Messerschmidt, *Die Wehrmacht im NS-Staat*, 447.

72. Berghahn, "NSDAP und 'Geistige Führung' der Wehrmacht, 1939–1943," 51.

73. Quinnett, "Hitler's Political Officers," 65.

74. *Zur Geschichte des Nationalsozialistische Führungs-Offiziere (NSFO),* 100–101. See also Zoepf, *Wehrmacht zwischen Tradition und Ideologie,* 126–127.

75. Quinnett, "Hitler's Political Officers," 94.

76. Messerschmidt, *Die Wehrmacht im NS-Staat*, 435.

77. Zoepf, *Wehrmacht zwischen Tradition und Ideologie,* 173, 179.

78. Quinnett, "Hitler's Political Officers," 40.

79. Zoepf, *Wehrmacht zwischen Tradition und Ideologie,* 203–205.

80. As cited in Messerschmidt, *Die Wehrmacht im NS-Staat*, 459.

81. Zoepf, *Wehrmacht zwischen Tradition und Ideologie,* 370.

82. Zoepf, *Wehrmacht zwischen Tradition und Ideologie,* 290

83. Based on a letter from Major Fellgiebel to Robert Quinnett, in Quinnett, *Hitler's Political Officers*, 223.

84. Quinnett, "Hitler's Political Officers," 261.

6

Political Officers in the Soviet Military

Regardless of what a political officer deals with, what question he resolves, he subordinates his work to one goal, one main task. That task is the same as the company commander—ensuring the company's constant high level of combat capability and readiness.

—N. I. Smorgo and P. F. Isakov

DESPITE SOME IMPORTANT MODIFICATIONS in the role played by the political officer/commissar during the 1930s and 1940s in the Soviet armed forces, the situation never deteriorated to the point where the political commissar controlled the actions of the commander. He would be reintroduced and he would play a pivotal role in political indoctrination and motivation just prior to and during the Second World War, but the commander would continue to enjoy considerable autonomy when dealing with military affairs. In the postwar period, the political officer was always clearly subordinated to the commander under the unity of command principle. In fact, in time the political officer developed into an important cog in the military machine, being responsible for matters such as morale, motivation, combat readiness, discipline, political socialization, and so on. The best evidence of just how important his role would become came in the last days of the USSR, as the country's military leaders pleaded with the country's political leadership to keep political officers and the Communist Party in the military because they served as unifying devices at a time when little else was left to hold the armed forces together, in light of the pressures for independence and autonomy on the part of the country's fifteen republics. In short, by 1989 the political officer was playing a crucial role in the military machine.

Background

Despite the change in functions that occurred with the end of the political com-missar position in the mid-1920s, the situation within the Red Army was far from stable. Beginning in 1937 the military underwent a major purge, as Stalin attacked the army—which he feared could and would turn on him in a crisis. On June 11, 1937, the media noted that certain military leaders had been arrested and that they would be put on trial on the next day. This group of officers included M. N. Tukhachevskiy, I. E. Yakir, and I. P. Uborevich. The first had been deputy people's commissar of defense, while the latter two were well-known commanders of mil-itary districts. These men and others were accused of a "breach of military duty and oath of allegiance, treason to their country, treason against the peoples of the USSR and treason against the workers' and peasants' Red Army."[1] The next day they were tried and executed. All of these men, but especially Tukhachevskiy, were prominent military thinkers and planners. All had played a major role in mod-ernizing the Red Army during the 1920s and 1930s, and all had at one time or an-other offended Stalin (often unintentionally, but that mattered little).

The purge that followed was as massive as it was complete. For example, one source noted that 3 out of the 5 marshals were shot, as well as 13 of 15 army com-manders; 57 of 85 corps commanders; 110 of 195 division commanders; 220 of 406 brigade commanders, all 11 vice commissars of war; and 75 of 80 members of the Supreme Military Council—including all district commanders. Altogether, this meant close to 90 percent of all generals and 80 percent of all colonels.[2]

What was even more important for the purposes of this study was that politi-cal officers figured prominently among the victims. Yan Gamarnik, the head of the Political Administration, committed suicide when it became obvious that his ar-rest was imminent. In addition, "His deputy, Bulin, together with most of the chiefs of the political administration in military districts, vanished." And the purge of the political officer corps, as well as other parts of the military, did not end there. In fact, according to Conquest, "Gamarnik's Political Administration suffered even more than the rest of the forces. At the top levels there was a clean sweep. . . . All seventeen Army Commissars went, with twenty-five of the twenty-eight Corps Commissars. At Brigade Commissar level, two survived out of thirty-six."[3] John Erickson estimated that somewhere between 15,000 and 30,000 regu-lar officers (out of a total of 75,000–82,000) were eliminated, while Conquest argued that "at least 20,000 political workers had gone under." This so decimated the political officer corps that "by 1938, more than one-third of all the Party po-litical workers had no political education at all."[4]

Political Commissars Are Reintroduced

Faced with the upheaval in the army and Stalin's always paranoid fear of a revolt, the Kremlin had to do something not only to maintain control over the military, but to ensure that it remained in a position to fight if that should be necessary. As a consequence, on May 8 Moscow introduced a number of measures. First, in all military districts, fleets, and armies, the military created a troika of three individuals to oversee matters. This troika was composed of the commander and two members of the military council. In addition, at lower levels the institution of the military commissar was reintroduced. An order dated June 7, 1937 relieved commanders of authority over political matters. The existing deputies for political work were replaced by military commissars. According to one key Soviet source, this was a temporary action, "taken because of concrete historical circumstances."[5]

One often misunderstood aspect of the decision to reintroduce commissars during the late 1930s has been the assumption on the part of many Westerners that commissars were duplicates of those in the 1920s. As one writer put it, "On May 10, 1937, the principle of one-man command instituted so recently in March 1934 was abolished and the dual command principle was re-established." Reading this writer's analysis, one comes away with the impression that the situation at that point was just like the one in the 1920s.[6]

In fact, there were significant differences. Most important, the political commissar did not have the kind of control over the line officer that he did during the civil war. "Military commissars . . . concentrated in their hands, all party-political leadership over soldiers, giving the commander the possibility to concentrate exclusively on the activities of soldiers, to broaden his knowledge, to raise combat readiness."[7] The main task of the commissar himself was political work—"The political commissar was recognized as the everyday leader of the political organs, and the party and Komsomol organizations."[8] The political commissar assumed authority for "party-political measures, ensuring that plans, combat tasks and the political preparation of the troops, and that agitation-propaganda, and cultural work are carried out."[9] The key point with regard to this arrangement, for our purposes, was that the military avoided a dual-command arrangement whereby both officers had to sign every order. Instead, authority was bifurcated: the commander had charge of military matters, while the political commissar was in charge of political affairs. The main difference between this and the previous situation was that the commander was not responsible for political matters as well as combat-related issues. For this reason, Soviet writers refer to the situation as the introduction of the "incomplete form of unity of command."[10] In practice, the arrangement was somewhat clumsy, with the lines of authority being vague. Each officer had responsibility for actions in his own area of competence, but if the commander was the senior officer, all orders were "issued in his name and in the first person."[11] Nevertheless, the situation was awkward at best and created problems, as the key Soviet source on this action admits, but given the way the purge had decimated both political and line officers, no other alternative existed. There was a severe lack of regular officers with the kind of military and political training necessary to enable them to assume responsibility for both functions. Similarly, given how unstable and volatile the situation was in the political realm as a result of the ongoing purge, there was a critical need for officers—many of whom were Communist Party members taken directly from civilian life—to push the party's line within the military. In addition, by separating the two functions, Stalin further lessened the chances that the military would work as a cohesive unit. This situation was not as divisive as it had been during the civil war, but neither was it as cohesive as it had been during the latter part of the 1920s and the early part of the 1930s.

Concerning the question of control, the role played by the commissars was different than it had been in the 1920s. As Petrov put it, "Some kind of control over the actions of the commander was completely excluded."[12] There was no procedure for having both officers sign orders.

Nevertheless, it would be naive to think that commissars did not have some impact on line officers. One of the political commissar's primary tasks was to check on the political reliability of all members of a unit—including, of course, the commander. Needless to say, the purge was still under way, and the concern for potential disloyalty was rampant everywhere. In addition, the commissar often doubled as party secretary, and in almost every case, an officer was first excluded from the party before he was arrested as an enemy of the people. A bad party report could be fatal. On the other hand, the political commissar also had to be careful of being denounced by a disgruntled line officer who was unhappy with the treatment he or others had received from the commissar. In any case, in those instances where questions arose about the political reliability of the unit's commander (or commissar), the matter had to be referred to more senior political organs, where a decision would be made concerning the individual's suitability for his post. And it should be remembered that political officers were purged almost as often as regular line officers.

Another factor that played an important role in the decision to reintroduce the commissar was the huge expansion that was under way within the Red Army. For example, in 1927 it was made up of 586,000 officers and men, but by 1937 the number had increased to 1,433,000.[13] This expansion required the addition of large numbers of officers and soldiers who were not party members. All of these soldiers, especially the officers, had to be carefully indoctrinated—not an easy task at a time when the purges were under way and everyone was uncertain what the party line was, because it could and often did change at a minute's notice. Nevertheless, from a party perspective, having officers whose primary purpose was political indoctrination meant that more time was available to inculcate both soldiers and officers with the regime's values.

As increasing numbers of soldiers fell victim to the purge, political commissars were also tasked to find new party members to replace those who had been purged. And the political commissars carried this out with considerable energy and success. For example, in 1937 alone, 13,200 new party members (and candidate members) entered the ranks of the Communist Party—raising the size of party membership in the military to a total of 147,500. The following year a total of 101,300 joined as members or candidate members. As a result, by the end of

1938, more than 200,000 members and candidate members of the Communist Party were in the military (a figure that did not include the 26,400 candidates and full members of the party in the navy).[14] The following year 165,000 new party members were recorded. These individuals had to be supervised (i.e., party organizations and structures had to supervise them). Thus, by the end of 1939 there were 9,468 primary party organizations (the smallest unit in the military) and a total of 435,000 party members. A party organization existed for every 50–60 communists. Clearly, the party organization was becoming a major tool of political socialization, playing an even more important role than it had in the past.[15] Furthermore, by making party membership almost mandatory for officers, the Kremlin was creating a situation in which professional soldiers were subject to two forms of discipline—one through the normal military chain of command and the second through the party. Even if an officer performed well in carrying out his military duties, he ran the risk of trouble if he failed to show the necessary enthusiasm in carrying out his party duties.

Because of the ambiguous nature of the relationship between the commander and his political commissar, problems soon began to develop. Who would be in charge in a crisis? How could one differentiate between political and military matters? As a result, in December the Central Committee issued a directive that called on party institutions (including the commissar) to work to strengthen the authority of both the unit commander and the commissar and to do everything possible to ensure that the two individuals worked well together.

Meanwhile, recognizing the need to improve the educational standards of political commissars, at the beginning of 1938 the party changed the name of the Political-Military Academy in Moscow to the Lenin Political-Military Academy. The number of students was increased three times relative to the preceding year, and the areas a student could specialize in were also increased. The quality of the faculty was upgraded and other institutions were created; for example, in 1940 the Kalinin Pedagogical Institute was established. In addition, 24 new political officer schools were created between 1938 and 1939. By August 1939, 6,000 students were studying at these political officer schools. By the first part of 1941, 10,800 political officers had graduated from them. In addition, by the beginning of 1941 30,000 additional officers had received political training from a variety of regular officer schools—which added political courses for their students.[16] Finally, assistant political workers were introduced. These soldiers were given the task of helping political commissars in carrying out their assignments. The obvious purpose behind these actions was to provide senior political officers with the kind of training that would enable them to understand party policy and to provide the necessary political leadership in the military. As a consequence, by the middle of 1940 more than 70,000 political workers were in the Red Army, three times more than in the previous three years. In practical terms this meant that by 1940, the number of empty political positions had decreased three times when compared with 1938.[17] Still, there were not enough political officers to go around. Furthermore, the qualifications of these political commissars left much to be desired, especially when it came to their knowledge of things like weapons, tactics, and equipment, not to mention morale and motivation.

The Soviets had known for quite some time that political officers would be more effective if they understood the weapons the soldiers used or the equipment they trained on. Nothing drew the scorn of soldiers more quickly than political commissars whose knowledge was limited to the elements of Marxist-Leninist theory. A way had to be found to train political commissars to handle—and master—the equipment and weapons the soldiers used every day. As a result, in March 1940 the Central Committee passed a resolution calling for better military training for political officers. During the year 1940 alone, some 40,000 reserve political officers received weapons and equipment training—a development that would have tremendous benefits for the army during the upcoming war.[18]

Meanwhile, a major effort was undertaken to provide the many soldiers who were being called up with a basic understanding of the regime's values. For example, journals and newspapers were created, which attempted to explain the relationship between the military and Marxism-Leninism, as well as the reasons why soldiers were expected to fight and, if necessary, die for their country. A new generation of lecturers, or propagandists, was prepared in order to ensure that the party's message was available at all levels. In addition, political commissars were made responsible for providing language classes for those soldiers who did not speak Russian—and there were a lot of them, especially soldiers from Central Asia! With the increasing numbers of non-Russians, political commissars were expected to help to overcome the deep-seated prejudices that many Russians felt vis-à-vis non-Russians. This led to the effort to push the idea of the "friendship of peoples."

In an effort to get out its message and to infuse ethnic unity and toleration, the party created 18 local military newspapers—in addition to the centrally controlled *Krasnaya zvezda*. By 1941 the number of newspapers put out by individual units was up to 500. Cultural clubs were also created, Lenin rooms were established in every unit (a total of 27,000 of them by 1940), and unit libraries became common. Most important—regardless of whether newspapers or Lenin rooms were involved, the political commissar was in charge of ensuring that everything functioned smoothly, of contributing to a high state of morale, and of politically motivating soldiers.[19]

Back to Political Officers

As the number of trained political commissars rose and as Stalin became less concerned about the military's reliability, the need for the inefficient and intrusive institution of political commissars correspondingly decreased. As one Soviet officer put it, "The high political level of commanders, their preparation for military as well as political leadership made possible the introduction of unity of command in all units and subunits."[20] As a result, on August 12, 1940, the Presidium of the USSR issued a decree on enhancing unity of command in the Red Army and Navy. This decree eliminated military commissars. Unit commanders now re-assumed full responsibility for all activities taking place in their units—including political education. To quote a Soviet source, "The Deputy Commander for Political Affairs, as the representative of the higher party organs, has direct responsibility for leading all party-political work under the supervision of political organs of units or the political apparatus of subunits."[21] The post of deputy commander for political affairs was introduced in all army and navy units, military training institutions, and other army and naval establishments. Unity of command was now back in full force.

In the meantime, efforts to increase party membership were intensified. During the first half of 1941, 29,700 soldiers became candidate members and 44,500 were accepted as full members. Thus by the time the Germans attacked the USSR, a total of 298,950 party members and 196,958 candidate members were in the ground forces. In addition, 41,442 party members and 26,153 candidates were in the navy, giving the military a total of 563,503 party members. Thus, 12.7 percent of all soldiers were party members, while in the navy the number stood at 20 percent.[22] There was little likelihood of a loss of party control or of the military being isolated from the party.

Then came Germany's attack on the Soviet Union on June 22, 1941. Despite the major increase in the size of the Red Army during the 1930s, there were not enough troops to meet the German onslaught. In the first days of the war, thousands of Soviet troops were either killed or taken prisoner. The USSR was in mortal danger. A State Defense Committee—under Stalin's leadership—was created to oversee the country's defense. The country's senior party and governmental organizations—the Central Committee and the Council of People's Commissars—issued a directive on June 29 that called for a restructuring of the party apparatus to meet the demands of war. Greater centralization of all party activities was critical. The USSR could only meet the German attack if all efforts were devoted to executing the war. Toward this end, the party took three significant steps. First, Communist Party workers at all levels were mobilized as political officers. In most cases these were full-time Communist Party workers—individuals who had spent most of their lives working within the civilian party machinery, in some cases overseeing agriculture, while in others working in industry, culture, or education. A sign of this mobilization of civilian party workers came in June 1941, when those attending Central Committee schools in Moscow and Leningrad found their course of instruction changed overnight to political-military affairs. "As a rule, the students who had worked previously as secretaries of regions or cities found themselves transformed into members of the armed forces." Altogether, some 25,000 party workers who had been attending party schools became political officers in the Red Army by the end of October.[23]

Second, all Communist Party members were called upon to join the army. Third, hundreds of thousands of young people were drafted into the army. In fact, given the extent of the threat facing Moscow, Russia needed not thousands but millions of young men. Past increases in manpower would pale in comparison with the new situation. By the end of 1941 60,000 party members and 40,000 candidate members had been mobilized—and in many cases sent straight to the front.[24] They were not well trained, but insofar as Moscow was concerned, they were better than nothing. Besides, the Kremlin hoped that their ideological zeal would help make up for what they lacked in technical training. And in many instances they were formed into special units because of their devotion to the party. In other cases, they were inserted in units where morale was a problem. Their task was simple: to raise morale by inspiring others to go above and beyond the call of duty in fighting the Germans.

Once Again, Political Commissars

Not surprisingly, a critical need arose for all kinds of officers to man this expanded military. Reserve officers were called up, but many of them had only received rudimentary training—if any at all. Furthermore, the fact that the country was now fighting a war enormously complicated the tasks faced by commanders. Given the chaotic situation, the need to rely on reserve officers, and the lack of military knowledge on the part of political workers, on July 16, 1941 the Kremlin again introduced the institution of military commissars in all regiments, divisions, headquarters, and military educational/training institutions. Military commissars also were added all the way down to company level. As far as commanders were concerned, the situation was similar to that in 1937. "It was clear that there were not enough qualified commanders capable of military and political leadership in the extremely difficult combat conditions."[25] As a result, military matters were once again placed in the hands of the commander, while the political commissar assumed responsibility for political questions.

In terms of training political commissars, the situation was better than it had been during the expansion in 1938–1939. For example, in the second half of 1942, the percentage of political commissars with a political-military education was 64 percent for infantry units, 81 percent in artillery units, and 96 percent in aviation units. Similarly, the percentage of political commissars who had been engaged in full-time political work prior to the start of the war stood at 74 percent in infantry units, 78 percent in artillery units, and 80 percent in aviation units.[26] Regarding the relationship between political and line officers,

> The military commissar is the representative of the party and of the government in the Red Army and together with the commander has full responsibility for the unit's carrying out of its military tasks, for steadfastness in combat, for unshakeable readi-

ness to fight to the last drop of blood against the enemies of our country and to not yield an inch of Soviet soil.[27]

As in the previous period, Soviet sources claim that the commissar had no control over the actions of the commander—as had occurred during the civil war.[28] His basic task was to relieve the line officer of the need to deal with political matters. "The political commissar was primarily focused on improving party-political work."[29] He was far too busy with working on military issues to worry about things like morale or political indoctrination. Concerning combat itself, tremendous attention was focused on the political commissar as an example of what a soldier should be. In contrast to the American chaplain, he was expected to lead his forces into battle. Indeed, Soviet literature on the Second World War is replete with stories about the courage of political officers—their willingness to brave the worst and most difficult situations.[30] Reading through the memoirs of both Russian and German participants of the early period of World War II, there is no doubt that the commissars did an outstanding job of strengthening the will of the average Soviet soldier. The troops' readiness to fight against almost impossible odds during that early period was due, at least in part, to the role played by the commissars.

One problem associated with the party's use of communists to shore up the army's morale and willingness to resist was that party members suffered disproportionately heavy casualties. Because they were in the forefront of the battle, large numbers were killed. As a result, the party was constantly working to recruit new party members. Toward this end, the military issued a directive that gave party membership to soldiers who had distinguished themselves in battle. Between August and December 1941, 126,625 soldiers joined the party.[31] From the commissar's standpoint, this meant even more pressure, both to maintain a lively party organization and to train new recruits in the basics of Marxism-Leninism.

One problem that continued to confront political workers throughout the Second World War was language. Too many soldiers, especially those from Central Asia, still did not know Russian well enough to carry out orders, not to mention to understand lectures on political issues or directions on how to operate equipment and weapons. In some instances, the Kremlin had set up ethnic units, believing that men of a single ethnic group would serve better together. The problem, however, was that the political commissar and the commander often had to work through an interpreter when they issued orders or gave political lectures. Often these lectures—and orders—were both poorly translated and not culturally sensitive. The main result was that both the commissar and his doctrine seemed to many soldiers to be foreign inventions: something forced on them from outside. Recognizing the extent of the problem, the army and the political administration went out of their way to provide newspapers in the local language—another task for the commissar. Efforts were even made to translate manuals into the local language, but in the end the language problem remained a major hurdle and headache for the commissars.

As far as the remainder of the army was concerned, all officers gradually began to participate in efforts to motivate soldiers for the difficult struggle against the Germans. Meetings somewhat reminiscent of those that occurred in Cromwell's army were held, in which all sorts of military leaders spoke. Commanders of forces at all levels, members of military councils, commanders and commissars from various units, as well as NCOs and average soldiers, gave speeches. Their purpose was to convince soldiers both that their cause was just and that their sacrifices would make the difference between victory and defeat in this increasingly bloody and unforgiving conflict. For example, discussions of German atrocities would be held in an effort to intensify soldiers' feelings of hatred toward the Germans. And this included not only discussions of actions against captured Russian soldiers but revelations of the Germans' slaughter of Russian civilians as well. Such meetings would also be used to give medals and other awards to soldiers who had distinguished themselves in battle.

If nothing else, these speeches, which became increasingly common in the middle of 1942, showed that the commanders were becoming more familiar with and more accepting of the party's line. Most of them were party members, a good source for commanders. Those whose political reliability the party questioned had long since been removed from command.

Political Officers Reappear

By the fall of 1942, it was becoming clear that the military was functioning smoothly, that the many civilians and reserve officers had been absorbed into the army and had proved that they were both willing and able to do their duty. In short, they had demonstrated that they were reliable. Similarly, regular officers were now familiar enough with the party's political line and had accepted it to such a degree that questions were being raised about the utility of the political commissar institution. Indeed, according to one Soviet source, it was becoming a "brake in the further development of leading soldiers."[32] Separating the political and military functions when it came to leadership within a unit was becoming dysfunctional and unnecessary.

There was a good reason why this relationship wasn't working. During the first two years of the war, the army was constantly on the defensive. In such a situation, it made sense to separate the political and military functions. The primary task of the former was to convince the troops—the majority of whom had been drafted—to defend the country, often against overwhelming force and odds. "Fight to the death" was the kind of slogan that was repeated over and over. Militarily, the task was rather simple. By the latter part of 1942, however, the situation had begun to change. The Soviet army's job now was to remove the Germans from Russian soil. In essence, the Red Army was increasingly going on a more complex offensive. And from a military standpoint, this required more initiative on the part of military commanders.[33] As a consequence, a mechanism was needed to assure that the commander had the flexibility to show initiative when that was necessary. This led to the issuance of a directive by the Central Committee on October 9, 1942, which again eliminated military commissars in favor of the "full unity of command."[34] And full unity of command was immediately granted to all commanders, regardless of whether or not they were party members. The commander now had complete responsibility for both military and political activities in the unit, although the political deputy continued to supervise the latter on a day-to-day basis. The important point, however, was that the political deputy worked under the supervision of and for the commander—even if he was not a party member. Political assistants remained in their positions. The only difference was that instead of working for political commissars, they were now working for political officers.

Insofar as the focus of attention of the new arrangement was concerned, Moscow made it clear that political work in the army was not to suffer. Indeed, all Soviet sources agree that the Kremlin expected political work to be even more intense. There was, however, an important change. The content of political work was to be expanded. Instead of focusing primarily on things like Marxism-Leninism, the pressure was now on to include a plethora of military-related subjects. In short, whatever helped advance the fight against Germany became grist for the political education mill. While some discussions might focus on traditional subjects like imperialism or German atrocities, now lectures began to deal with military issues as well—for example, to inspire troops prior to an upcoming battle, push the need to master new and different weapons, convince them to learn new kinds of tactics, or discuss how to treat German prisoners of war. And the Soviets were not just talking about the changed relationship between the commissar and the commander. In addition, the commander was expected to work closely with the party organization (most units had their own party organization). What was most interesting about this development, however, was that it was made very clear to the party structure that its primary function was to ensure that the commander had all of the support he needed to carry out his military tasks. "The transition to full unity of command demanded that the commander make skilled use of the party organization, and mobilize the powers of communists in carrying out concrete military training and educational tasks."[35] That did not mean, however, that all commanders willingly made use of political officers—certain commanders wanted nothing to do with them or at least did not make effective use of either political officers or the party organization.

The task of both a political commissar and the deputy for political affairs was not an easy one during World War II. To begin with, as noted in the last chapter, Hitler issued the so-called Commissar Order, which called on German troops to kill—without mercy—every political commissar they captured. This was partly a result of the hatred Hitler felt vis-à-vis the Bolsheviks, and it was partly a recognition on the part of the Germans that such individuals played an important role in helping to motivate Soviet troops. The political officer or commissar was also expected constantly to set an example—to be the first in the fight. If an attack was ordered, the political representative was expected to be in the front lines, fighting and inspiring his troops. If the troops did not fight hard, or if they were not maintaining their equipment the way the commander believed they should, the political representatives would be called on the carpet to explain the problem.

> Together with providing political education for the Red Army soldier, they were supposed to be a personal example of fearlessness and combat courage, of inspiration to the troops. In that sense, especially high demands were placed on the company political officer. The first in the attack, the last to relax—he was obligated to combine his personal example with daily educational work.[36]

Not surprisingly, the losses suffered by political workers in the Second World War were enormous. During the first months of the war, for example, political commissars had among the highest casualty rates in the Red Army. In some units, political commissars were replaced three or four times—in just a few months of action. There were some 50,000 dead among political commissars in units formed prior to February 1942. And most of them were in senior positions—as company political officers or as staff officers at higher levels.[37]

As a result, the party went out of its way to recruit new political cadres. For example, by the second half of 1941 there were more than 250,000 political workers in the Red Army—almost three times more than when the war began.[38] The Red Army's political training schools were overtaxed in their effort to provide the needed political workers. By the end of 1941 there were fourteen political-military training centers that prepared political officers for higher level positions. Seventy-six officer schools prepared young men to become political workers. Interestingly, the course content in these schools differed significantly from what had been taught prior to the war.

The country was now in the midst of a major war. Courses had to be shortened in order to get as many men through them and to the front as possible. Furthermore, instead of soldiers spending a lot of time on political questions such as Marxism-Leninism, greater effort was devoted to military topics. After all, these men were expected to go directly into combat and it mattered little if they understood the intricacies of Marxism-Leninism but could not fire a weapon accurately. Such a situation was a recipe for disaster. The political worker would be killed before he could make any kind of contribution to the unit to which he was assigned. This does not mean that political factors were ignored—the political worker was still responsible for motivating his troops! Rather, it meant that the main emphasis would be on matters of direct relevance to winning the war—and in this case, that meant how to utilize weapons and fight.

In the first six months of the war, these schools turned out a total of 57,000 political commissars.[39] They were not sufficient, however, especially in light of the tremendous number of casualties that political workers had suffered. In fact, some 10,000 political workers would die during the war. As the situation worsened, training courses were introduced at the battalion and division level to train junior political officers. Individuals who had an interest in becoming political officers or those whom the party selected would be sent to these short-term training courses and, after completing them, would go to the front.

While it may seem insignificant to civilians, one thing that had irritated political workers most was that their rank structure was different from that of line officers. This rank difference only emphasized the commonly held belief that political workers were not "real" military officers. This situation changed in October 1942, however, when political workers were given military ranks just like their regular officer counterparts—but only after they had proven to an examination board that they were capable of performing the military tasks expected from a line officer of the same rank.

Equally important was the decision taken in December 1941 to provide political workers with an opportunity to become line officers. As in the past, they had to have served for at least three months in combat, had to prove themselves, and had to pass muster in front of an examination board, but given their experiences under fire and the crying need for more line officers, many were accepted into the ranks of regular infantry, tank, or armor officers. By October 1942 more than 4,500 political workers had become commanders, and by the end of the war approximately 300 commanders of regiments were former political workers. Overall, by the end of the war, 150,000 former political workers had become commanders of regular army units at various levels.[40]

Faced with the large number of political workers that had transferred to the ranks of regular officers, the Kremlin decided to cut back on political officers by combining many positions within the ranks of the political apparatus. As a result, the number of officers (especially at senior levels) in the political apparatus shrunk by more than half. The reasoning was simple. The unity of command principle had been introduced and the commander was now in charge of political work. Furthermore, the ideological mettle of thousands of men had been tested in battle. They had proved themselves to be loyal, dedicated communists and soldiers. The need for officers who spent most of their time working to motivate these individuals had lessened significantly. As a consequence, the number of schools and educational institutions being used to train political officers also decreased to the point that by the end of the war, only 43 such institutions existed in the Soviet armed forces.[41]

By 1943 pressure to reorient political officers' training to incorporate more technology had intensified. Although there were fewer of these officers, most were unfamiliar with the basics of military technology—despite the Kremlin's efforts to provide for a more balanced education. The demand for such officers had been so great that their training had been cut short, and for practical purposes they knew almost nothing about military matters.

As a consequence, in 1943 the military high command issued an order that ended the lengthy four-year program that trained senior political officers. The course of study at the Lenin Political-Military Academy was immediately reduced to one year. Similarly, an order was issued stating that all political workers should be provided with basic military training, and in March Stalin signed an order that listed the minimum military knowledge required of a political worker for promotion. As a result, all of them had to pass an examination the following November. In a certain sense, this turned the tables on the political officers. Instead of having them lead political discussions that the rest of the unit was obligated to listen to, political officers now had to attend classes put on by their colleagues, on basic military matters ranging from how to fire a weapon to how to read a map. When the time came, the majority passed the exam. However, this was only a recognition that they knew the minimum necessary. As a consequence, these military courses (some of which lasted three to ten days) continued throughout the course of the war, in an effort to help political officers gain the stature and respect that soldiers would only give to officers whom they considered competent in military matters.[42]

Another way the Kremlin tried to help political officers carry out their tasks was by expanding the party organization. The idea was that instead of putting everything on the commander's or the political officer's shoulders, part of the burden would be transferred to the unit party organization. In essence, this meant that if minor discipline problems arose, the party organization would step in to try and deal with them. Similarly, if the problem was one of motivation—perhaps a soldier was not mastering the new technology fast enough—then the party organization would dispatch someone to deal with him. The same was true of personal problems. A soldier who was suffering from the death of a loved one clearly needed additional support—and this could be provided by the party organization.

Thus in 1942 a party organization was created in every regiment, every battalion, and other smaller units. By the end of the first six months of 1942, a total of 10,000 party organizations were present at the regimental level on the Western Front. Within the Red Army as a whole, the number stood at 87,485.[43] It would be hard to imagine a situation in which a soldier would not have contact with the party organization, particularly considering that by the middle of 1942, a total of 1,413,870 Communist Party members were in the Red Army. And as the situation facing the Red Army (e.g., around Stalingrad) worsened, party membership rose further. By the end of 1942 it was up to 1,939,227 and by the end of 1944 it was 3,030,775—the high point for the entire war.[44] Party membership was becoming a symbol—a sign that the individual concerned was dedicated fully to defeating the Germans. Furthermore, the ubiquitous presence of the party meant that almost all aspects of a soldier's life came under official or semi-official scrutiny. This strengthened party presence also placed an obligation on it and its leaders. If the unit was not performing at the necessary level, the party organization, as well as the commander and political officer, would be called to task. Why had the party not done what was necessary to ensure that everything was being done to achieve its military and political objectives?

Political officers underwent a number of transformations during the period just prior to and just after the start of World War II. They had problems—many were hardly qualified for the tasks they assumed. The majority came from civilian life or were soldiers who were pressed into becoming political workers. Their technical knowledge was never as high as it should have been, but from all appearances they met the test. They suffered heavily in terms of casualties and won their share of medals for bravery. Coming out of the Second World War, however, it was clear to everyone that political officers lacked the kind of training and expertise necessary to function in the postwar Red Army. Indeed, once the postwar downsizing of the military was finished, a lot of work would have to be done to upgrade the quality of the political officers. The task would be made even more difficult because many officers who had shown the most bravery under fire were the least qualified to function in the increasingly technical postwar Soviet armed forces.

The Postwar Political Officer

The postwar Soviet military had three basic tasks. The first was to consider the lessons from the war. And the Soviet military did this in a much more systematic manner than is normal among Western armed forces. Up to the present day, Russian military intellectuals continue to look closely at the events of the Second World War, with an eye toward understanding the process of combat better—and using those lessons to modify how wars will be fought in the future.

The second task—and the one that really hit political officers hard—was to absorb and master the new military technology that would clearly play an increasingly important role in the future. The third task was closely associated with upcoming the Cold War. The Soviets knew that a major struggle with the West was on the horizon and, as a consequence, worked hard to keep soldiers highly motivated and their military in a ready state. Combat readiness would take on a new sense of urgency.

> In the interest of defending our country, the difficult international conditions, the arms race in the main capitalist countries, demand from commanders, political organs and party organizations that in the future they will perfect combat readiness, strengthen military discipline among the troops and educate them in the spirit of devotion to the fatherland, to the communist party.[45]

In an effort to make certain that the party organs were taking the appropriate steps, a group of party inspectors was created. The task of these officers was to make sure that political officers and the party organizations were fulfilling their duties. In addition, political directorates were set up for the main branches of the armed services. This need for more supervisory structures was caused by the exposure of many Soviet soldiers to the West. Many of them were perplexed at the high living standard in almost all of these countries—a quality of life that far exceeded anything they had seen in the USSR. How to understand or explain such a situation? Furthermore, a lot of officers had enjoyed far more autonomy and had been given far more initiative than was normal for the Red Army during the war. In addition, a number of individuals who had become party members during the heady days of the war were not the kind of politically disciplined, committed individuals that the party wanted over the long haul. These "war-time" communists would have to be weeded out. The situation in Eastern Europe had to be explained to the troops and discipline had to be reimposed. The inspectors were there to make sure that both commanders and political officers were doing their jobs.

Needless to say, a huge educational undertaking was in order. Toward this end, the Marxist-Leninist Evening University was created. Such courses had existed in the past, but they had been discontinued during the war for lack of time and resources. Now officers would be expected to attend these classes—which at this point were almost entirely political in nature. As a unifying symbol, primary attention was devoted to Lenin himself. Anything that could be found in his writings to justify current Soviet policy was more than helpful. These classes—which the political officer and the commander were expected to attend—lasted two hours every month. Middle-level officers would be divided into another group, which would meet for a total of ninety-six hours per year, while the junior ones were ordered to attend "commander's lectures." The commander himself was expected to lead or at least participate in such lectures.[46] At the same time, considerable effort was put into improving military newspapers. By the end of 1946 there were 580 newspapers for soldiers. Many of them were intended for the members of a particular unit, but they often reprinted major articles appearing in central newspapers such as *Krasnaya zvezda*.

Meanwhile, political officer qualifications left much to be desired. As the primary Soviet source on the issue noted, "In accordance with the new tasks it was necessary to revise the system of political officer cadres of Soviet Army and Fleet."[47] A new system of education was needed, and it included making significant revisions not only in entry-level schools (i.e., officer schools) but in the educational system for middle-level and senior officers as well. In short, major changes in the way political officers were trained were in order.

The issue was simple. About 73,500 political officers were in the Red Army. Almost all of them had extensive combat experience. There was no question about their military qualifications. They had proven that they could meet that challenge. The problem, however, was that in spite of their practical experience, their knowledge of ideology was superficial at best. More than half of the political officers at battalion level had graduated only from a short-term course or a shortened program at an officer's school. A fourth of all political officers at the battalion level had not completed a military-political educational institution. Those serving in higher positions were also unqualified—after all, any political officer serving in a responsible position at division or regimental level should have completed the Lenin Political-Military Academy and very few had even gone through the short-ened courses as a result of the war. In August 1946 the Central Committee ordered the military to fix the problem.[48]

The first thing the high command did was to expand the capacity of the Lenin Political-Military Academy. A four-year course of study was set up, one that cov-ered areas such as the military press (i.e., training military journalists to become editors of the many military newspapers and journals that existed throughout the country) and economics. Second, the size of the faculty was expanded until it was twice as large as it had been previously; the academy was expected to become "an important center of scientific military-theoretical thought."[49] The quality of in-struction at the Kalinin Military-Pedagogical Institute was expanded as well—in fact, a special course of study in political indoctrination was added to better pre-pare political officers to work as indoctrinators. Changes were also made at middle-level and officers' schools. For example, instead of only a few months' re-quirement for someone to be declared a political officer, the course was length-ened to two years. In some cases, line officers were sent to such schools and upon graduation they became full-fledged political officers.

In time, this investment in education began to pay off. By 1953, for example, one-third of all political officers had a middle or advanced political-military edu-cation. Progress was especially evident at the regimental level and above. By 1952, 43 percent of all deputy commanders at those levels had an advanced higher mil-itary political education—eight times more than in the past. And most political officers (93 percent) were under forty-five.[50] By 1959 four-fifths of all political of-ficers had either a middle or an advanced political military education. One year later the percentage of political officers with advanced training at the regimental level and higher was up to 70. In this sense, they were better educated than most of their line counterparts.[51] The Soviet military had succeeded in retraining the majority of its political officers, while at the same time ensuring that the military did not get top-heavy from an age standpoint.

To further strengthen the role of political officers, a deputy for political affairs was added at the company level in 1950–1951. During the first stages, the deputies were included in units outside the boundaries of the USSR and in Western mili-tary regions. During the second stage, they were added to the remaining units. The reason for the presence of officers at this level was simple. The Cold War was on, and this was the level at which interaction between soldiers and their officers was most common and widespread. Moscow wanted to make sure that its message was getting through to the average soldier, and this was the perfect way to achieve that end. Furthermore, since political officers were also responsible for discipline, they would be better able to get a feel for what was going on among the troops. This arrangement was to last until August 1955, when political officers were removed from the company level.

The Political Officer up to 1989

It would not take long, however, before the high command began to realize that political officers were important factors in the maintenance of a high level of combat readiness and that such officers had to be much better trained than in the past. After all, if political officers were going to make a serious contribution to the unit's combat readiness, they had to have the respect of their line contemporaries. That would never happen as long as they were looked upon as "party hacks," officers unable to perform even the most basic military tasks.

Toward this end, two important steps were taken in 1967. First, the two-year political officer schools that had existed prior to 1959 were replaced with a new educational system. (Between 1959 and 1967, individuals becoming political officers were given short-term training courses and then sent to the higher-level political academies once they were senior enough.) These new schools had a four-year curriculum. In a certain sense, they were similar to American military academies. The courses during the first two years were almost the same as those of regular officer candidates. Prospective political officers studied weapons, strategy, tactics, topography, communications, gunnery, and other military-related matters. They also took courses in mathematics, physics, electronics, applied mechanics, and engineering draftsmanship. In their last two years of school, they focused on politically related issues or the "social sciences." They also received practical training as political officers in line units—similar to the internships, summer camps, or summer cruises aboard navy ships common to the American military.

In addition, the political officer schools were specialized. For example, there were schools for political officers serving on submarines, on surface ships, in communications, in armor divisions, or in the infantry. The idea was to provide these young men with the kind of technical training that would get them respect from line officers, as well as from the troops they worked with.

This raises a key question. What did political officers do on a daily basis during the last thirty years of the USSR's existence? Soviet sources dealing with the postwar period seem pretty much in agreement on this topic.[52]

First, and foremost, the primary function of any political officer was the "maintenance of a high level of combat readiness."[53] At almost every opportunity, Soviet sources make it clear that whatever else a political officer did, if the unit was not prepared for combat, the officer was not doing his job. For example, if the soldiers were not cleaning their weapons properly, then the political officer had failed to motivate them.

Second, the political officer—together with the commander—was responsible for morale. If there was a problem with morale, then the political officer was not doing his job. Was the food all right? Were the quarters satisfactory? What about personal problems? If a problem arose, it was up to the deputy for political affairs to bring the matter up with the commander and see that it was corrected. In this sense, the Soviet political officer carried out many of the duties normally assigned to chaplains in the American military.

Discipline was another area where the deputy for political affairs continued to play a key role. If discipline was lax and there were problems with soldiers' behavior, it was up to the political officer to correct the situation. Soldiers who had discipline problems were not being properly led. The political officer might decide to call on the party organization to help out, or, if the matter was serious enough, he could turn to the commanding officer. In any case, he was expected to deal forcefully and effectively with the matter.

As the world became more high tech, the political officer was increasingly saddled with the task of convincing soldiers of the need to master military specialties. For example, the Soviet military had three levels of military specialists (third class, second class, and first class). These specialties were split according to weapons branch. For example, there were three levels of specialists for tanks. If a soldier did not possess the third-class medal, it was up to the political officer to make certain that he was hard at work on it, and if he had the second-class medal, he should be hard at work on obtaining the first-class medal. Or what about the physical fitness medal? All soldiers were expected to be working hard to achieve and maintain it, since physical fitness was a crucial factor in sustaining a high level of combat readiness.

The deputy for political work was also responsible for party work. It was up to him to ensure that materials were available for the Marxist-Leninist University, to watch the status of the Lenin room, to deliver political indoctrination talks, and to discuss the importance of good relations between various ethnic groups. Lest the reader get the wrong impression, one political officer did not take care of all these political functions alone. Much depended on his rank and where he was assigned. For example, the company-level political officer was alone and pretty much in charge of all aspects of political work, while at higher levels one officer would be assigned to oversee youth (Komsomol) work at lower levels, while another might be in charge of agitation and propaganda. My point is that regardless of where he was assigned, the political officer was involved to some degree in political matters.

This brings us to the question of control. All Soviet sources are unanimous in stating that at no time during this period did political officers have the task of "controlling" the actions of line officers. Sources freely admit that this was the main task for political commissars during the 1920s, but they maintain that this did not happen when commissars were reintroduced in the 1930s and 1940s—although they do admit that confusion resulted in many cases because the lines of authority were blurred. Regarding the post–World War II political officer, all Soviet sources argue that this was never a serious issue. The unity of command principle meant that the commander was in charge—period!

This does not mean that political officers could not cause problems for their commanders if they decided to do so. They had access to a second chain of command—through the party—and could use it to complain about line officers. In addition, they had to fill out periodic reports on the reliability of regular officers. The problem with making use of either vehicle was that it could be seen as a failure on the part of the political officer to carry out his job. The commander would have to be pretty far out of line for a political officer to report him. If the political officer was wrong, there would be serious retribution. I have met more than one Soviet political and line officer who noted that reports became very routine over time. "So and So is a reliable officer fully devoted to the Party" became standard language—to the point that senior political officers would often complain that the reports were too superficial.

In trying to come up with a word to describe political officers, I was struck by the phrase that Herbert Goldhammer used: an expert in "human management."[54] If one were to compare the political officer's role to a corresponding position in the American military, he would be a combination of a chaplain, an education officer, and an executive officer. The comparison is not exact, but it illustrates just how multifaceted his role was.

The Value of the Political Officer to the Military

Political officers often served as lightening rods in the Soviet military—as vehicles for introducing and advocating whatever new policy the military leadership decided on. For example, when Gorbachev came to power, he pushed the policy of perestroika, or restructuring. The generals were fully aware that a major revolution was necessary within the armed forces, if the Soviet military was to be competitive in the twentieth century. The biggest problem facing the high command was how to convince officers—especially line officers—not to fight the last war but to focus on the next one.

To illustrate: As early as 1971 Marshal Ogarkov blasted Soviet officers for their failure to look to the future. "There is . . . I believe a more substantial reason for the shortcomings pointed out. It lies in the fact that individual military leaders do not keep pace with life and the development of scientific thought."[55] For his part, the late Marshal Akhromeyev, who played such a key role during the Gorbachev era, stated that history "offers many examples when the armies of some states prepared for future wars, basing themselves only on the experience of the past without taking into account the evolution of the military field."[56]

The initial reaction on the part of Moscow's senior generals vis-à-vis perestroika was hesitant. They gave lip service, but nothing seemed to change. When it came to making the point within the military, it was General Lizichev, who was head of the Main Political Administration, who picked up the ball.

> Even the election-and-report meetings are far from taking a demanding look at perestroika, from achieving fully collective work in the search for new forms and methods in effectively resolving tasks. In some places, criticism carries a formal, superficial character. At many meetings, criteria characteristic of bygone days, an insufficiently fresh form of analysis, and a lack of sharp conclusions and self-criticism predominate.[57]

And the political administration was not only assigned the task of waking up the rank and file of the military to the importance of perestroika, it was also given the responsibility of seeing that it was implemented. Marshal Yazov, who was defense minister at that time, noted "that each leading officer steadfastly implements the policy of the party, relies on the party organization, goes all out to support and develop an active and enterprising approach among communist and the entire personnel, and to guide them in resolving the task at hand." Or, as he described the role he expected the political apparatus to play later in the article,

> Being in the vanguard means always being in the midst of the mass of servicemen, knowing their sentiments, responding to the most burning problems in life of military collectives, and doing everything to resolve them, fighting persistently to assert a Leninist style of work, a creative businesslike style to achieve high final results.[58]

Yazov's emphasis on the role of the party in carrying out perestroika helps explain why the generals would later become so concerned about attempts by the political leadership to abolish the party. The party and political officers fulfilled an important function: they not only assisted in maintaining discipline, they served as a transmission belt between the military leadership and the rank and file. When the generals decided to implement a new personnel policy, the party-political apparatus could always be relied upon to help in implementing it. After all, most Soviet officers were members of the party, so they were subject to party, as well as military, discipline.

By February 1990 the party's presence in the military was under strong attack. A plenum of the Central Committee was held February 5–7 to draft a platform for the Twenty-Eighth Party Congress to be held later that year. In looking at the role to be played by the party in the military, the plenum adopted a thesis that called for a renunciation of the CPSU's monopoly of power, guaranteed by Article 6 of the Soviet Constitution. The plenum also called for the creation of a new Soviet presidency. The change was ratified by the Third Congress of People's Deputies in March.

This was not good news for the generals. General Mikhail Moiseyev, the chief of the General Staff, had publicly opposed the removal of Article 6 from the constitution. After noting that neither he nor any other military deputies had been permitted to speak at the plenum, Moiseyev observed that "many questions . . . arise in connection with Article 6 of the USSR Constitution, and in particular with the role of political organizers in these circumstances."[59] The ambiguity of the language included in the party platform was an obvious concern to the chief of staff. From his standpoint, removal of the party from the army was only a matter of time.

A political officer interviewed at the end of February provided an interesting alternative commentary on the role of the party in the military. In response to a question concerning his views on the Article 6 controversy, he replied that he "thought the article must be reviewed." This individual said he believed that the party must provide "a guiding and directing role not declaratory, but on the basis of some sort of legislative acts and its work in society, also including the Armed Forces." He warned that the party and the political organs must remain in the military. "With all the critical attitude toward the party apparatus and toward the party leadership today, this apparatus is the only real structure of civilian authority and it is a consolidating organization."[60] This officer had hit the generals' main concern on the head. Given all the pressures that were trying to pull the military apart, the generals realized that the party organization and the political officers were two of the few structures holding it together.

Both the generals and the high command admitted that steps needed to be taken to improve the quality of political work—work by political officers. The commander of the Moscow Military District remarked that the role of the political organs (of which political officers were a part) needed to be "enhanced," while at the same time he admitted that there were shortcomings in their operation. We must be sure that "they are constantly and closely concerned with men's lives and can contribute to their civic and military training," he added.[61] All kinds of suggestions were being made on how to improve the role played by political officers. General Vladimirov, for example, argued that political officers should be trained in the social sciences.[62] He also suggested the need to create a Directorate of Education and Culture in the Ministry of Defense to take over the functions of the Main Political Administration. Its tasks, according to Vladimirov, would be:

> to educate servicemen in a spirit of Soviet patriotism and internationalism and propagandize civil law; make personnel aware of modern knowledge and achievements of science in the sphere of psychology, political science, sociology, and other social sciences, and the values of Soviet and world culture; protect servicemen's social and political rights; make contacts by Army and Navy units and formations with public organizations and the mass media; and work constantly and purposefully to elevate the prestige of service in the USSR Armed Forces.[63]

This approach was heading in the right direction. However, from the vantage point of the generals, the author failed to include issues like discipline and interethnic disputes. Who would keep the military together if Gorbachev disbanded the party political organs and sent political officers packing?

The Twenty-Eighth Party Congress in effect removed any party influence from the political structures and made them part of the armed forces. The generals now had complete control over the political organs. This new situation was formalized in a decree that Gorbachev signed on January 11, 1991. This document formally made all political structures directly subordinate to the Ministry of Defense. In noting the primary tasks of the political structures, the document focused on areas such as discipline, education, morale, social justice, and work with civilian organizations. The military really did not have time to work out the details because the coup in August led to a collapse of the Soviet Union and the Communist Party. Political officers were soon out of work, although the Russian Army took some of them over as educational officers. The experiment with political commissars, and later political officers, was at an end.

Conclusion

If anything is evident from the evolution of the role of the political officer in the former Soviet Union, it is that institutions of this type change over time. From political officer to political commissar to political officer and then back to political commissar. This was followed by the reintroduction of political officers. But even this was not enough. The role played by political officers changed over time. For most of the postwar period they were used to help maintain a high level of combat readiness in the military. Then Gorbachev took over and they began to serve as cheerleaders and the primary force behind the effort to introduce perestroika in the armed forces.

What was most surprising—in light of the general view on the part of many Westerners—was how important political officers were to the country's generals. This does not mean that the generals always liked them or that there were no differences of opinion. I have spent considerable time with Soviet political and line officers. A Soviet admiral, who was in charge of submarines in the Northern Fleet, commented to me that "these people can't even stand watch." He was concerned that a political officer was taking up valuable space on a crowded submarine but could not carry out the necessary technical functions expected of every officer. It is also worth noting that Russian military officers often referred to political officers as "staff rats" (*Shtabnaya krysa*) because they almost never held command positions. On the other hand, I have witnessed cases where a group of line officers was celebrating the selection of one of their group to attend the Political-Military Academy and become a political officer.

Concerning the question of control, this was not the task of either political commissars or political officers during the period covered by this chapter. Secret police officers who worried about such things were present in the military (as in all sensitive institutions). Similarly, although a political officer could turn one of his regular counterparts in for lack of party dedication, this seldom occurred. Indeed, a number of officers told me privately that senior political officers often complained that the reports political officers wrote on their line colleagues took on a formalism that said little about the real situation; for example, "Major P is a reliable and dedicated communist" seems to have been the standard phrase, as noted earlier.

Political officers in the Soviet military played an important—if sometimes changing—role. They were crucial during the Second World War, and they fulfilled a useful function in the postwar period. The best evidence of how important they were came in the closing days of the USSR, when the high command fought

hard to keep political officers in the armed forces. They are now history, but someone had to worry about things like discipline, morale, motivation, and political
socialization. In that sense, they played an important, even pivotal, role.

Notes

1. *Pravda*, 11 June 1937, as cited in Robert Conquest, *The Great Terror* (New York:
Macmillan, 1968), 202.
2. Leonard Schapiro, *The Communist Party of the Soviet Union* (New York: Random
House, 1959), 420.
3. Conquest, *The Great Terror*, 227–228.
4. John Erickson, *The Soviet High Command* (London: Macmillan, 1962), 505, 506.
Conquest, *The Great Terror*, 228.
5. Yu. P. Petrov, *Stroitel'stvo politorganov, partiynykh I komsomol'skikh organizatsiy
armii I flota (1918–1968)* (Moscow: Voennoe Izdatel'stvo, 1968), 238.
6. Michael J. Deane, *Political Control of the Soviet Armed Forces* (New York: Crane,
Russak, 1977), 41–42. One gets the same impression for other writers. For example,
Schapiro, *The Communist Party of the Soviet Union*, 421; Erickson, *The Soviet High
Command*, 478.
7. Petrov, *Stroitel'stvo politorganov*, 289.
8. Petrov, *Stroitel'stvo politorganov*, 290.
9. Petrov, *Stroitel'stvo politorganov*, 239.
10. Petrov, *Stroitel'stvo politorganov*, 239.
11. Alexander Khmel, *Partiyno-politicheskaya rabota v sovetskikh vooruzhennykh silakh*
(Moscow: Voennoe Izdatel'stvo, 1968), 28, 29.
12. Khmel, *Partiyno-politicheskaya rabota*, 239.
13. Khmel, *Partiyno-politicheskaya rabota*, 28.
14. Petrov, *Stroitel'stvo politorganov*, 242, 244.
15. Petrov, *Stroitel'stvo politorganov*, 246.
16. Petrov, *Stroitel'stvo politorganov*, 248.
17. Petrov, *Stroitel'stvo politorganov*, 250.
18. Petrov, *Stroitel'stvo politorganov*, 251.
19. Petrov, *Stroitel'stvo politorganov*, 266.
20. Petrov, *Stroitel'stvo politorganov*, 261.
21. Petrov, *Stroitel'stvo politorganov*, 262.
22. Petrov, *Stroitel'stvo politorganov*, 264.
23. Petrov, *Stroitel'stvo politorganov*, 277.
24. Petrov, *Stroitel'stvo politorganov*, 278.
25. Petrov, *Stroitel'stvo politorganov*, 287.
26. A. M. Batov, *Deyatel'nost' Kommunisticheskoy partii po podgotovke I osushch-
estvleniyu edinonachaliya v period Velikoy Otechestvennoy voyny* (Moscow: Voennoe Izdatel'stvo, 1966), 105–106.
27. *KPSS o Vooruzhenykh Silakh Sovetskogo Soyuza* (Moscow), 360.
28. Petrov, *Stroitel'stvo politorganov*, 289.
29. Petrov, *Stroitel'stvo politorganov*, 290.

30. See, for example, G. M. Mironov, *Komissary na linii ognya* (Moscow: Politizdat, 1984).

31. Petrov, *Stroitel'stvo politorganov*, 295.

32. Petrov, *Stroitel'stvo politorganov*, 310.

33. The Russian approach to war has never permitted the kind of initiative that soldiers in the West are familiar with. Actions at lower levels are almost always approved by more senior officers. Nevertheless, in actual combat, offensive actions require the commander on the spot to show more initiative than in a defensive operation.

34. Petrov, *Stroitel'stvo politorganov*, 310.

35. It is worth noting that political commissars remained a part of the landscape in partisan units until January 1943. Fighting battles behind enemy lines against an unforgiving enemy like the Germans, with the need to deal with civilians on a widespread basis, was quite different from running a highly disciplined military unit. Petrov, *Stroitel'stvo politorganov*, 311–312.

36. Petrov, *Stroitel'stvo politorganov*, 315.

37. Petrov, *Stroitel'stvo politorganov*, 316.

38. Petrov, *Stroitel'stvo politorganov*, 316.

39. Petrov, *Stroitel'stvo politorganov*, 317.

40. Petrov, *Stroitel'stvo politorganov*, 319.

41. Petrov, *Stroitel'stvo politorganov*, 321.

42. Petrov, *Stroitel'stvo politorganov*, 323.

43. Petrov, *Stroitel'stvo politorganov*, 328–329.

44. Petrov, *Stroitel'stvo politorganov*, 329, 337.

45. Petrov, *Stroitel'stvo politorganov*, 391.

46. Petrov, *Stroitel'stvo politorganov*, 397, 398.

47. Petrov, *Stroitel'stvo politorganov*, 400.

48. Petrov, *Stroitel'stvo politorganov*, 400–401.

49. Petrov, *Stroitel'stvo politorganov*, 402–403.

50. Petrov, *Stroitel'stvo politorganov*, 405.

51. Petrov, *Stroitel'stvo politorganov*, 513.

52. Among the sources consulted for this section were M. Butskiy, I. S. Mareev, and A. S. Skachkov, *Politiko-vospital'naya rabota v podrazdelenni* (Moscow: Voenizdat, 1982); M. G. Sobolev, *Partiyno-politicheskaya rabota v sovestkoy armii i flota* (Moscow: Voennoe Izdatel'stvo, 1984); N. I. Smorigo and P. F. Isakov, *Zamestitel' komandira roty (batarei) po politchasti* (Moscow: Voennoe Izdatel'stvo, 1982); V. I. Komissarov, *Polevaya vyuchka i politrabota;* and I. S. Mareev, *Partiyno-politicheskaya rabota v Sovetskoy Armii i Flote* (Moscow: Voennoe Izdatel'stvo, 1967).

53. Mareev, *Partiyno-politicheskaya rabota v Sovetskoy Armii i Flote,* 74.

54. Herbert Goldhammer, *The Soviet Soldier* (New York: Crane, Russak, 1975), 277.

55. N. V. Ogarkov, "Teoreticheskiy arsenal voennogo rukovoditelya," *Krasnaya zvezda,* 10 September 1971.

56. S. Akhromeyev, "Prevoskhodstvo sovetskoy voennoy auki I sovetskogo voennogo iskusstva—odin iz vazhneyshikh faktorov pobedy v velikoy otechestvennoy voyne," *Kommunist* 3 (1985): 50, 62.

57. "Ostree osenivat' reshitel'no deystvovat," *Krasnaya zvezda,* 15 November 1986.

58. Y. D. Yazov, "Perestroika v rabote voennykh kadrov," *Kommunist vooruzhennykh sil* 7 (1987): 4, 10.

59. "Zadachi u nas odin," *Krasnaya zvezda*, 10 February 1989.
60. "Kakoy jbyt' armii 90-kh," *Argumenty I fakty*, 24 February 1990.
61. "Aktivnost' mysli I deystviya," *Krasnaya zvezda*, 13 March 1990.
62. "Voennaya reforma: opyt, problemy, perspektivy," *Voennaya mysl'* 5 (May 1990): 50.
63. "Voennaya reforma: opyt, problemy, perspektivy," 50.

7

Political Officers in the East German Army

As is well known, political work is conducted to make the army into a reliable
instrument of the particular state and the policies it follows.

—Colonel Eberhard Haueis

IF ANYTHING, THE EAST GERMAN case confirms the Soviet practice. In the beginning, when there was a serious problem with control, political commissars were
utilized. As soon as that problem passed, however, political officers quickly took
their place. Indeed, for most of the existence of the East German military/police—
and certainly during the life of the National People's Army (Nationale Volksarmee
or NVA)—the political officer served as the commander's deputy and had very little influence as a "controller." He shared with the commander responsibility for
the unit's combat readiness, which included not only morale and motivation but
political reliability as well. Should the unit not pass the appropriate tests or should
an investigation show that combat readiness was not at the required level, the political officer faced serious consequences. A soldier was not only expected to love
and serve the German Democratic Republic (GDR), he was also expected to be a
first-class soldier. In short, the political officer was responsible for motivating
troops when it came to political issues, as well as for pushing them to achieve
higher levels of military competency.

The Main Administration for Training

When the German Army collapsed at the end of World War II, the country was
without a military. In the Eastern sectors of the country, the void was filled by the

Soviet armed forces, which occupied the area. Recognizing the problems involved in administering the region, Soviet military authorities began to set up German police units. Their task was to carry out the orders given by the Soviet high command, as well as to maintain public order. They were expressly forbidden, however, from carrying out any kind of military exercises. This was purely a police force. Then, a Ministry of the Interior was created (DvDI) and by November 1945 approximately one hundred Germans were being utilized by the Russians to secure the occupation zone's borders.

Not until the Foreign Minister's Conference in April 1947, however, did the Soviets decide to give German officials authority over police matters. By this point, it was becoming clear that both sides—East and the West—would go their separate ways in dealing with their zones of occupation in Germany. As a consequence, while the Russians retained control over weapons, munitions, intelligence, and the actual issuance of orders, the East Germans were given authority in other areas. For example, on June 1, 1947, a Conference of Interior Ministers (from the different German lands or states) accepted responsibility for fighting espionage, as well as for maintaining public safety.

One thing that is important to keep in mind is that the primary criteria for selecting a cadre to man these police forces—even at this early date—was political reliability, a factor that took priority over technical expertise.[1] It did not matter how good a police officer was at his job—if the Kremlin could not be certain of his loyalty, he was of no use to Soviet authorities.

In 1948, East German party leader Walter Ulbricht informed the Interior Ministry that it was time to form a *kasernierte*[2] police force as a means of improving the coordination and operation of the police in the Soviet Occupation Zone. One key problem, however, was that there very few ideologically committed young men who could be taken into the new military. As a consequence, a way had to be found both to socialize these young men into the value system of the new communist system, as well as to ensure that they and any other recruits obeyed their ideological leaders in East Berlin.

Utilizing Former Members of the Wehrmacht

Another problem arose for the East German communist leaders. They could not simply ignore the absence of military expertise on the part of the members of the *kasernierte* police. There were very few trained police officers among the ranks of the country's communists. Furthermore, East Berlin had more in mind than just better trained police units. Communist leaders were beginning to think about the possibility of military-type units. From the East German perspective, military forces would help move the Soviet Occupation Zone, or SBZ, closer to the status of a sovereign state, something that East German party leaders desperately wanted.

The only reliable sources of manpower available to East German military authorities were either those who had fought in the Spanish Civil War or the few German communists who had some military experience, in a few cases in the Russian armed forces. As a consequence, the East Germans had no alternative but to rely on former Wehrmacht officers and soldiers because of their military expertise.

It was absolutely necessary to take a number of military specialists into the police, that meant officers and even very high ranking officers because we didn't have such specialists in our party, who had the proper training in weapons, the tactical training and above all the provision of supplies which a supply officer is able to provide under all conditions. For that reason, we found it necessary from the beginning to adopt a course of accepting certain specialists in these areas.[3]

The East Germans were embarrassed to have to take this step, but they could do nothing else about it. They had to have the expertise.

Yet these individuals—some of whom had been in Russian captivity—could hardly be considered reliable. Many of them were offered early release from Russian prison camps if they would agree to serve in the new police units. Not surprisingly, many jumped at the chance. Indeed, as early as 1947, former members of the Wehrmacht were in police units. One source claims that they made up 5.6 percent of all officers and 28 percent of all noncommissioned officers (NCOs).[4] Some of these individuals would occupy senior ranks in the *kasernierte* police forces. On October 1, 1948, some one hundred officers and five generals who had served in the Wehrmacht returned from the USSR to take up their positions as officers in the police forces.[5] By the middle of 1951, a third of all officers had served in the Wehrmacht. All together, some 431 officers, 956 NCOs, and 2,004 enlisted men who had previously been members of the Wehrmacht were serving in the East German police forces.[6]

The former Wehrmacht officers were not the only problem. The average soldier knew little or nothing about the regime he was supposedly serving. Even worse, he was unfamiliar with the communist ideology that made it different from others. A way had to be found to indoctrinate soldiers in the ways of the new regime.

Creating a Political Apparatus

Given the close ties between the East German Communist Party and the Soviets (one could argue that without the Red Army, there never would have been an East German state or a functioning Communist Party), it is not surprising that East Berlin turned to the Soviet experience for inspiration. The result was the creation of an East German version of the Soviet political commissar, the so-called PK officer (Polit-Kultur Offiziere). As Ulbricht put it,

We must popularize the People's Police and ensure that they have reached a level where they understand our policies. For this reason it has been decided that a political cultural leader will always stand along side the leader in the police. He will counter sign the order and be personally responsible for the political and ideological condition of every party member, and ensure that there is a unitary party organization within the ready units (*Bereitschaften*) of the police. These organizations are not sub-ordinate to the local party organizations or those in the Kreis or in the Land, but on the contrary, they are subordinate to our directives and discipline.[7]

From a theoretical standpoint, these new officers, who were introduced in the spring of 1948, were in a strong position. No line officer could issue an order un-less his PK officer countersigned it. The underlying concept was that no action by the police could be considered separate from its political implications. To quote one writer on the subject, "Without the agreement of the PK-leader the orders and instructions of the commanders and leaders were invalid."[8] Hoffmann[9] described the party control function and the power of orders of the first deputy with the as-sertion that it was up to them to "stimulate and politically carry out the overall work of the People's Police." He based this comment on the premise that there can be no "separation between political and substantive-police work."[10]

Needless to say, not only the former Wehrmacht officers were not prepared for the new world they encountered; many political officers were in just as bad shape. For example, Werner Rothe, who was to serve for many years as a political officer in the East German military, noted in his biography that when he was appointed a political officer in 1948, he had no idea what a political officer was supposed to do. As far as he could determine, his main qualification was that he was a mem-ber of the Communist Party. Rothe was assigned to an officer's school where he was the PK officer. His primary task seemed to be to ensure that the students did well. He did not give political indoctrination lectures—these would not begin until 1950. His other tasks included: ordering newspapers, getting tickets for cul-tural events, setting up a library, organizing celebrations and social events, and making sure that birthdays were observed.[11]

Toward the end of 1948, GDR party official Wilhelm Pieck met with Stalin. It was clear that while the Russians were happy with efforts by East German officials to begin to build the core of a military, they did not want to do anything that would make the eventual goal of German reunification impossible. Stalin still be-lieved that a neutralized Germany was at least a possibility. As a result, all of the efforts the East Germans were making to build an East German military machine were taken in great secrecy.

The Main Administration for Training

Not surprisingly, the creation of the PK officer led to many of the same problems that the Soviet military had experienced in its early days. The PK officer would eventually be assigned ideological work, and during the early years of the Main Administration for Training (HVA), about 30–35 percent of training time was de-voted to ideological work—after all, few of the troops knew much about commu-nism. In addition, regardless of how reliable the commander might be, this fledgling political officer also had to sign all orders—even, which was often the case, when he knew little or nothing about the military issues at hand. How could a commander be expected to respect his PK officer when the latter didn't know his left foot from his right foot? Most PK officers had come directly from civilian life and then were sent to PK officer school. This training was not much of a help because 75–90 percent of their time was spent on ideological courses. As a result, there was constant conflict between PK officers and their line colleagues. Indeed, it led to a situation where by the end of 1952, of the 800 PK officers who had been trained in 1951, 300 had to be released from service as unsuitable—unable to function effectively in a military environment. Berlin tried to improve the situation by requiring that PK officer candidates have at least one year's service as an officer or enlisted man in the *kasernierte* police units prior to attending the officer's school.[12] But the basic problem would not go away.

Political reliability remained a major problem in the HVA. In January 1949, Order No. 2 was issued. This order called for a major purge in the organization. In the beginning it focused primarily on the border troops because of the political sensitivity of their positions, but it was gradually expanded to include other troops as well. Among the criteria that were used to select individuals to be purged were the following:

- soldiers who had direct relatives in the West,
- those who had been interned by Western forces after the war,
- those who had received numerous disciplinary citations,
- those who had shown themselves to have bad character or who were morally unreliable, and/or
- re-settlers and those who appeared to be politically unreliable.[13]

As a result, about a third of all border troops were released from duty. The same fate would befall a number of regular police forces. By the end of the year, however, the order had lost its validity.

One continuing problem facing East Germany's political and military leaders was recruiting soldiers and officers for the HVA. This is why so much pressure was put on German prisoners of war, who were being held in the Soviet Union after the war, to join the HVA. In Germany itself, pacifistic tendencies were evident in the population as a whole, and in young people in particular. The country had just been through a major war, and few men looked forward to the idea of putting on a uniform. Indeed, they had been told ever since the fall of the Third Reich that German militarism was one of the worst fates to befall Europe in the last several hundred years. Why join the German military? To deal with this situation, East German officials not only pressured party members to join, but they made a lot of promises to attract young men. The problem, in the first instance, was that even party officials (who were in practice drafted) often had little interest in the armed forces. As a result, the promises they made were generally not fulfilled.[14]

Major morale problems also plagued the troops. Food was often bad, quarters were unlivable, and all types of clothing, equipment, and weapons were lacking. Things were so confused that in the beginning, training units were using World War II German training manuals as well as Wehrmacht weapons.

In short, the situation within the HVA was close to chaotic. Moscow did not want the West to know what was going on—although it was not really a secret—while at the same time, the East German party leadership wanted desperately to field an army, a modern, highly trained, and equipped military that would serve to further the SBZ's claim to sovereignty, something the East Germans were trying to convince the Soviets was in their own interest. It would be a long and hard struggle, but the idea of an independent, even if only a partly sovereign, state was always in the back of the minds of East Germany's leaders. In fact, throughout the country's existence, its leaders would constantly worry that the Soviets would sell out to Bonn in order to get something from the West Germans. To counter this prospect, East Berlin's leaders did everything possible to make themselves indispensable in the eyes of the Soviets. They tried so hard that many in the West thought they were being paranoid. In the long term, however, their worst fears would be realized—the Kremlin would sell them out in an effort to get West German aid and assistance.

Meanwhile, recognizing the seriousness of the problems facing the HVA, East Berlin undertook a significant reorganization—one that lasted from the end of October 1949 to November of the following year. The old structures were dissolved as the organization was placed directly under the control of the Ministry of Interior. Then, the two types of units—the *Bereitschaften,* or police units, and the educational establishments—were unified and placed under the Ministry of Interior. Summer camps—to train personnel in military-related skills—were also added, and the *Bereitschaften* were restructured along the lines of military units. Then new, more German-oriented uniforms were introduced. In fact, according to one observer, these changes had little impact on the troops themselves—with the exception of the new uniforms and the decision to release those soldiers who either had close relatives in the West or who had been in Western captivity after the war.[15]

At the same time, another major purge was carried out. For example, between September 1949 and October 1950, 10,000 men were released because of concerns about their political reliability.[16] Then, in an effort to improve discipline, a basic organization of the Free German Youth (FDJ), the party's youth organization, was introduced and in December the first conference of the FDJ was held.

The situation facing the HVA's leadership was not an easy one. Desertions were a constant problem, and the level of political loyalty was open to question. For example, only 24 percent of all officers, NCOs, and soldiers were party members—a figure that would contrast with almost 100 percent of officers and 50 percent of NCOs in later years.[17] Then there was the continuing problem with pacifism. Far too many young men continued to feel that they had had enough with war and war games. Nevertheless, in spite of these problems, the party forged ahead. A PK school was set up in Torgau to prepare officers to act as the deputy PK, a PK teacher, a propagandist, or an FDJ secretary, as well as to train officers in cultural affairs and mass agitation work.

Equally important, in an effort to win the loyalty of those who would serve in the armed forces, the military gave a clear preference to recruiting individuals with proletarian backgrounds. By 1953, for example, 65 percent of HVA members came from the working class. Sixteen percent were from the countryside, with the remainder spread among a variety of occupational backgrounds.[18]

In the meantime, considerable effort was devoted to building up a party organization. By the end of 1949 party organization had begun to stabilize. Members were increasingly called upon to help with ideological training and, most important, to serve as an example for their nonparty colleagues. This was especially evident in July 1950, when the party called on PK organs and the party organization itself to do more to make all members of the HVA into conscientious defenders of the first "Peaceful German State."[19] At the end of April 1950 the highly competent Heinz Hoffmann, who would later lead the NVA, took over as head of the HVA. By the end of that month, the size of the HVA was up to 51,548.[20]

Despite these efforts, the situation within the HVA continued to be unsatisfactory. For example, at the Second Party Conference of the SED (Sozialistische Einheitspartei Deutschlands) in July 1950, speakers blasted the HVA for both its political and its military shortcomings. PK functionaries and the party organization were called upon to study the conclusions reached at the Party Conference and to show how they planned to overcome these difficulties. And regarding ultimate responsibility for dealing with these problems—especially the political ones—it was made clear that the political apparatus bore primary responsibility.[21] In a direct effort to improve the situation, all HVA officers were told that if they wanted to remain in the police, they had to give up their membership in one of the "bloc" parties and join the SED (the Communist Party). A few members of the NPDP (National-Demokratische Partei), which primarily had been set up for former members of the Wehrmacht, were allowed to remain non-SED party members, but, by and large, by the end of the year all officers were SED members.[22]

Another sign of the increasing politicization of the HVA was the form of address that was used. In the early years, *Kamerad*, a term that had little or no political connotations, was employed. That was replaced by the term *Herr* until 1952, when the very political term *Genosse* was introduced.[23]

Soviet Advisers

Given the low level of technical competency on the part of most members of the HVA, not to mention serious questions concerning their political reliability, it was not surprising that Soviet advisers were widely utilized. They were already present in the *Bereitschaften* in July 1949. On September 15, 1949, all of the schools received Soviet advisers with the rank of VP (*Volkspolizei*) Inspector. The same was true of the police units, which also received a *VP Oberrat*. All of these Soviet officers wore HVA uniforms to disguise their presence. At the same time, the first selections were made for long-term training for future East German military officers in the Soviet Union.[24] This would become a very important factor in training East German officers—including political officers. Indeed, it would later be almost impossible for an officer to make flag rank if he did not attend a Soviet military academy. According to one source, between 1950 and 1989 about 13,500 East German officers were trained in the former USSR. That number included 385 officers and generals who had completed the General Staff Academy.[25]

By 1950 approximately 25 to 30 Soviet advisers were present in every regiment. Companies had one Soviet adviser, while at the battalion level there was also a Soviet adviser. Three or four Soviet officers were present on the regimental staff, while the remainder were distributed throughout the units—primarily as technical advisers. As far as their relationship to the East German commander was concerned, "The Soviets were present in the *Bereitschaften* because they really were individuals with outstanding knowledge and good understanding. While they were supposed to help with the leadership of the *Bereitschaft*, technically, they never issued orders—do this and do that!"[26] It is hard to believe that Soviet advisers never interfered in the command process in practice. In fact, their influence was often indirect. This is clear, for example, from the advice that Wilhelm Zaisser (at that time head of the HVA) gave his officers: "One would always be wise to follow the advice of the Soviet adviser, although in the last instance, the *Bereitschaft* commander should always decide what is good or right for a *Bereitschaft*."[27] Or as another source put it, "It was therefore demanded that officers 'accept with open hearts' the advice given by their Soviet advisers."[28] One could always complain about a Soviet adviser, but it was clear that such complaints were not welcome. By and large, however, East German officers, including political officers, seem to feel that they played a positive role. To quote one officer who went on to become a career political officer in the NVA, "What I especially valued on the part of the Soviet officers was that they were not pedantic and impatient with us young officers."[29] Almost all of these Soviet officers had fought in the war and had lost a good many members, if not all, of their families. Yet, according to this writer, they played a very positive role in spite of what they had seen and suffered at the hands of German troops during World War II.[30] While they continued to serve in the NVA as advisers until its demise, over time the number of Soviet advisers would decrease to the point where they were only present at the highest levels.

Politicizing the Military

Returning to the question of PK officers, it soon became obvious that most regular officers, as well as political officers, believed that a clear distribution of labor

existed. It was the line officer's task to run the military, not to worry about politics. As a result, most line officers were incapable of answering even the most basic
political questions—for example, why is the Soviet Occupation Zone for peace
while the West is preparing an aggressive war? PK officers believed that their only
obligation was to carry out political work, not to worry about technical matters
such as weapons, equipment, or tactics. Such an approach was unacceptable to
party leaders.

As a consequence, the party took a couple of steps. First, it was again made
clear to line officers that apolitical behavior would not be tolerated. To drive that
point home, the purges continued. In 1950 a total of 198 HVA members were
ousted from the party—and in practical terms, that meant from the military. Political schooling was also intensified—indeed, during the early 1950s, 30 percent
of all training time was devoted to ideological training.[31] Second, the PK apparatus set up a long-term training program. Not surprisingly, there were problems.
How to handle the Wehrmacht experience? Some PK officers went out of their
way to avoid discussing this topic, thus leaving officers unsure of how to deal
with their past. On the other hand, there were those who tried to transplant the
Wehrmacht experience—after all, this was what they knew and understood. The
simple fact was that most young PK officers were overtaxed both intellectually
and physically. Finally, the work of the party organization was also stepped up.
New party members became increasingly involved in supervising the behavior of
members of the HVA.

Meanwhile, in a desperate effort to improve the qualifications of the PK officers, a major educational effort was undertaken. Up to 1950, future political officers spent a year in school and were then commissioned and sent to a unit. About
90 percent of their training was on political subjects. With the reorganization of
the HVA, the course of study was lengthened to two years.[32] Even then, PK officers' training was almost entirely in the political area. Another former political officer estimated that almost 70 percent of their time was devoted to political subjects. PK officers were being trained to carry out tasks such as a weekly two-hour
session of political work (with different classes for NCOs and enlisted personnel),
organizing and carrying out party and FDJ work, engaging in personal discussions with soldiers concerning their individual problems, and making daily com-
mentaries on the world's events.[33] Part of the problem, however, was a lack of
guidance—too often, no one in East Berlin knew the party's line on the various issues. Herbert Peter, who was himself a political officer at the time, noted that he
often received daily phone calls from authorities in Berlin, telling him what he
should say to the troops about issues such as West German rearmament.[34] Obviously, a PK corps that had to rely on daily guidance was not in a position to inspire a sense of ideological confidence in its troops.

What to do? The answer came in 1952, when the party decided to introduce
the "Single Leadership Principle" (*Einzelleitung*). Instead of making both officers equal, the party now seemed confident enough to permit the line officer to

become the commander—in all areas—just as had happened in the Soviet Army once Moscow decided that the military was reliable enough to modify the old dual authority structure.

The introduction of the single leadership principle would revolutionize the roles of both the commander and the PK Officer. The commander was made responsible for all activities within the unit he was commanding. This meant political and military work. This new arrangement impacted heavily on the commander as well as on other line officers. No longer could either individual escape involvement in political work. If political work was not successful in an officer's unit, he would receive an unsatisfactory fitness report. At the same time, this new approach also had serious implications for the PK officer. First, as in the Soviet Army, his title was changed to deputy for political work. Second, he could no longer hide behind his political work. He was just as responsible for the unit's military performance as was the commander. The same was true of its political performance, morale, and motivation. In fact, from this day forward, both officers would increasingly find themselves in the same boat. If the unit failed to carry out its training missions, then this was just as much the political officer's responsibility for not motivating the unit as it was the commander's for not developing the right kind of training plan. The same was true of desertion. Obviously, the commander was responsible—but so was the political officer. How could such a thing happen if the political officer was on top of matters and knew the situation with regard to each individual in the unit? In short, it became the political officer's job to ensure that the principle of unity of command was a reality. As one writer put it, "The PK officer is responsible for the overall political level of the *Bereitschaft*. This does not relieve the commander of his responsibility. The commander is the chief, the PK officer his deputy. The PK officer must always keep in mind that how 18-20 year old individuals develop depends on his influence."[35]

This brings us to the political officer's job at this point in time. To begin with, all political officers had to be SED party members. They were expected to ensure that both military training and ideological education courses were successful. They would be called upon on a weekly basis to provide ideological lectures to NCOs and enlisted personnel. In addition, they were expected to carry out a whole series of other tasks. For example, Godau reported that he was told to watch over construction material while the unit was building its barracks, to ensure that it was not stolen. To make matters even more interesting, he was given the task of keeping order when a group of girls showed up at the construction site to help build the barracks! The girls were to be kept away from the soldiers at night—a task that he found very challenging. Then, on another occasion, he was told to accompany his unit commander as they went from one town to another in an effort to recruit personnel for their understaffed unit.[36]

In terms of ideology, the political officer was expected to explain to the troops—and officers—the intricacies of Marxist-Leninist theory, as well as to convince the many pacifists of the need for a modern, efficient military. To help him, the army employed "agitators," enlisted personnel whose task it was to find interesting articles and to report on them to the troops. The agitators also helped the PK officer by bringing problems to his attention—for example, if an individual was having a hard time accepting the party's view of things, or if an individual had serious personal issues. The political officer could only intervene if he knew such problems existed.

The PK officer's biggest challenge, however, was to explain the importance of close ties between East Germany and the USSR, as well as why Soviet culture was superior. Having been only recently conquered by the Russians and having experienced firsthand the excesses of Soviet occupation, many soldiers still retained a strong feeling of anti-Sovietism. How to instill a love of the USSR and all things Russian when many of those who listened to the political officer actually loathed the Soviets?

Finally, it was also not surprising that a lot of pressure was put on the political officer to push members of his unit to improve their qualifications in the military sphere. At the most basic level, the political officer was responsible for discipline and technical competence. He was expected to motivate personnel and ensure that their morale was at the highest level possible, as well as take good care of the technology and weapons systems. "World political questions were sometimes passed on as information, but we did not have major, strategic discussions. For us it was always the concrete results of military practice."[37]

In an effort to further tighten up political control in the HVA, in July 1952 an independent Political Administration was created. It was in charge of all political work in the HVA—including separate entities for naval and air units. Instead of PK officers having to call East Berlin for guidance on what to say every day, this administration was tasked with making up such material in advance. Similarly, the administration was expected to prepare lesson plans on how to motivate soldiers, train them, and discipline them. In addition, basic organizations—the core unit of communist organizations—were created at the regimental, battalion, and independent company level. In the latter instance, the purpose was to strengthen the political officer's hand by giving him additional resources to help him in the political socialization process.

No issue was more serious at this time than discipline. From an ideological standpoint, the problem was obvious. In the Wehrmacht, iron discipline had been the rule of the day. Now, however, the party faced a different situation. It still wanted iron discipline, but party officials had to argue that conditions were very different. After all, that was one of the critical points they were trying to make to their soldiers. In the old fascist army, soldiers were exploited by the "evil" ruling class. Now—in the communist view—a new political system had been created, one that served the interests of the overwhelming majority of the East German populace. The trick was to convince the average East German soldier that he had to submit to the same kind of brutal discipline, but, because of the change in political systems, it was in his interest to do so.

Despite the party's best efforts, the situation was far from satisfactory. For example, desertions continued to be a problem. In 1951, 395 members of the HVA deserted to the West. In 1952 the number rose to 1,218, including 7 PK officers.[38] What to do? Again the answer was simple. Make the PK officer directly responsible, because "desertions were a reflection of the overall unsatisfactory political moral situation including a lack of vigilance as well as a continuing lack of ideological influence."[39]

The KVP Is Created

In the meantime, Stalin had begun to realize that the division of Germany was becoming permanent—that the chances for a united, albeit neutral, Germany had decreased to the point where it was time to move on, from a policy standpoint. As a consequence, East Berlin was given permission to take the next step in the establishment of an army—the creation of the *Kasernierte Volkspolizei* or KVP. This was still a half measure, however. Moscow did not want openly to sanction the creation of an armed force, so the Kremlin continued with the subterfuge that it was a police force—even though no one, including members of the KVP, had any doubts about East Berlin's intention.

On July 1, 1952, the KVP became a reality. In addition to ground troops, it also included a separate naval and air arm. Heinz Hoffmann became the head of the KVP, while Willi Stoph was made interior minister (in accordance with East Germany's public denial that it was building a modern army, the country did not have a Defense Ministry), Wlademar Verner became head of the Navy, and Heinz Keßler took over as head of the air arm. By December, the KVP's membership was up to 90,250.[40]

Regarding the political situation, on July 17, 1952 the political organs were restructured. Main Administrations for Political Affairs were set up in the navy and the air arms. This permitted the Main Administration to provide the army, navy, and air force political administrations with information on a weekly basis. Unfortunately, this did not solve the problem. The material given to the political officers was very general, and the officer himself was required to provide examples. Few officers had the kind of educational background that would enable them to play this role effectively. The bottom line was that the political officer did not understand enough about communism, military technology, or the area in which the unit was stationed to resolve the many issues that cropped up daily. Questions were a constant problem—the lecture lasted for fifty minutes, but the course went on for two hours. Many of these political officers laid awake at night wondering how they would be able to deal with the questions that the soldiers inevitably asked during this period—for example, "Why are we not permitted to travel to the West?"; "Why is life in the West better than in the East?"; "Why can't we listen to Western radios?" Another problem facing the political officer was the need to work with the ten or twelve agitators in the unit. If the questions from the general population were difficult to deal with, the situation was even worse when it came to the agitators, most of whom were either party members or candidate party members. Their questions were even more demanding and complex; for example, "How does the dialectic explain the superiority of Western technology? If the USSR is more advanced ideologically than the West, why is the living standard in the West higher than that in the East?"

One major problem facing these political officers—and a situation that would haunt them until the end of the GDR—was that they did not command much respect on the part of their line colleagues. Political officers were often looked upon as nothing more than political indoctrinators in uniform. After all, their major function at this point in time was to teach soldiers and officers enough ideology so that they would understand the basics of Marxism-Leninism. Furthermore, it was not so long ago that political officers had enjoyed joint command authority, a situation that the line officers had never found acceptable. Those political officers who had a military background made out somewhat better. Hagermann, who was himself a political officer at this time, noted that the fact that he had graduated from artillery school and had spent two years as a regular line officer put him in good standing with his commanding officer. The latter had considerably more confidence in Hagermann because of his military background.[41]

This lack of military background was a problem when the political officer attempted to carry out one of his main jobs—motivating the troops to carry out their military tasks in an outstanding manner. He could always come up with catchy phrases like "Every target hit—a victory over the opponent," but his comments would have far more value if he himself was able to fire at a target and score a hit every time.

To help him in his job, the political officer had books, films, and visits from outsiders—a situation that was reminiscent of the French Revolution. Old-time communists or individuals who had suffered at the hands of the Nazis during World War II were often brought in to talk to the troops, to try and convince them of the superiority of the new system and to give them reasons why they should be willing to lay their lives on the line for the new socialist state.

Needless to say, the task facing the political officer was not an easy one. To begin with, there were not enough of them. In January 1952 there were 1,556 political officers, but by the end of that year, the party estimated that it would need 4,641 of them. Ninety-eight percent of those who became political officers came from industry or agriculture—these men did not have the best educational background. By the end of 1952 only a total of fifteen political officers had some form of higher education. Eighty percent were under twenty-five. Most political officers who joined the military did so in an effort to raise their social and economic status.[42]

By 1953 the political situation within the armed forces was getting worse. In fact, it reached the point where the SED leadership found it necessary to publish "an open letter to members and candidates of SED in the armed forces." Even more pressure was placed on commanders and political officers to ensure that the soldiers and officers were both reliable and technically competent. A report issued in April, however, indicated that the situation continued to deteriorate. "The reports confirmed that the previous situation in the political-cultural situation continued to exist. . . . The fact is that while in some units conditions have improved, overall the situation has not changed."[43]

As a consequence, the next month a statute on the "Organs for Political Work" was issued. For the first time, the functions and organization of the political apparatus within the military were clearly delineated. The political organs were given a special educational role within the armed forces. Their primary functions were: first, to raise the level of combat readiness, and second, to convince soldiers to hate communism's enemies while at the same time instilling in them a love for socialism and the socialist fatherland. The FDJ was also given enhanced status and made directly subordinate to the SED party organization. As one source put it, a soldier who possessed a high political-moral level was one who would give "unconditional obedience, endure extreme physical endurance, and possess outstanding military knowledge."[44]

The Berlin Uprising of July 17, 1953 showed just how tenuous East Berlin's hold was on the KVP. To begin with, there was considerable hesitation on the part of the country's political authorities to use the KVP to put down the uprising. Who would the soldiers shoot at? The demonstrators? Their officers? The political leadership? No one knew for sure. The average KVP soldier did not know what was going on. That this dissatisfaction was felt by a good part of the populace was clear to everyone. At the same time, as in the USSR, military personnel were cut off from the rest of society. They were separated from the civilian world when it came to work, and the little time they had available to themselves tended to be spent with other members of the armed forces. In short, they did not know how bad the economic situation really was.

As far as their actual use was concerned, military personnel were placed in a heightened state of readiness and used primarily to guard key public buildings or agricultural entities. They were told that weapons would be used only in extreme cases. When it came to crowds, only bayonets were to be used.[45] The real problem was that the KVP was still in the early stages of building political reliability. The KVP was certainly more politicized than the civilian population, but it would have been silly not to assume that each unit had been influenced by its surroundings. After all, if the 1953 uprising proved anything, it was that the leadership did not enjoy wide support.

While the KVP was not directly involved in putting down the revolt—that task was left primarily to Soviet troops—the revolt had an inevitable impact. First, the growth of the KVP was slowed considerably. Previously, the Russians and their East German friends had looked forward to building a very large military— 300,000 to 400,000. However, under pressure from Moscow the size of the KVP was reduced to 99,784 by September. By the end of the year, the KVP was down to

90,250. In 1953 alone, almost 3,500 officers were released for various reasons—most prominently, because of their assumed lack of political reliability.[46] The Russians made their dissatisfaction clear with the KVP and the East Germans in general, by withdrawing some modern weapons they were planning to provide this fledgling army. For example, General Baarß reports that 100 MIG 15s were sitting at Cottbus ready for the East German pilots to be trained to fly them. As soon as the uprising took place, the Russians flew them back to Russia because they were worried about what the East Germans might do if they got their hands on the MIG 15s—perhaps fly them to the West![47]

In order to explain the uprising, the party predictably laid the blame on West German intelligence or the CIA—a line that political officers were expected to parrot, even when they knew that was not the reason for the unrest. When something went wrong, it was always assumed to be a result of poor political work—a lack of training or education. What to do? Intensify political work![48] Furthermore, Soviet advisers—who were in the process of being withdrawn—reappeared throughout the KVP. Moscow was not taking any chances. Political reliability would figure very prominently for the next few months. After a year, these advisers were gradually withdrawn, first at the company level and then at the battalion and regimental levels. They remained at the division level.[49]

For its part, the East German leadership intensified control over the cadres. For example, by the end of the year, the Cadre Administration had been expanded to sixty officers. All three generals assigned to the Cadre Administration were old-time communists. They and those working for them had full control over all aspects of personnel.[50] In addition, the process of *Nomenklatura* (which meant that important military positions had to be approved by a party body) had been introduced. This ensured that the Politburo of the SED would have to approve all senior military appointments. Other positions—for example, those below general but still considered sensitive—would have to be approved by other bodies. The point was that East Berlin was determined to make certain that the officer corps was completely reliable. Regardless of how competent an officer might be from a military-technical standpoint, he was useless if he was not completely reliable from a political standpoint. Finally, a collegium was created within the KVP. This body's task was to monitor carefully the political reliability of the KVP. In particular, it was to look at questions such as: How well are the party's directives being carried out? How well is the military performing in a variety of areas, both military and political? How effective is political training?

In a nutshell, the main task of the political organs and political officers up to 1955 was to create an effective and politically reliable military force. This meant not only coming up with an effective means for getting rid of the "expert only" attitude prevalent among a lot of line officers, it also required the newly formed political officer corps and their supervisors to come up with a means for turning military personnel into reliable, but also technically qualified, servants of the East German state. Studying the Soviet experience was important—but only to a degree. These soldiers were not Russians but Germans, and to be effective, political education had to be made culturally relevant. The happy medium that such an approach required could only be found by trial and error. In addition, political authorities had to continue their efforts to produce a politically—and militarily—qualified group of political officers. They were improving—as the yearly report of the KVP noted in 1955—but political officers still had a long way to go before they would win the respect they needed from their line colleagues if they hoped to be effective.[51]

Meanwhile, the military continued to experience difficulties in attracting enough men to fill its ranks (almost half of its billets were empty). For that reason, in 1955 the party put out a call for party members to do their duty by enlisting in the KVP. The party suggested (there was no conscription at this time) that all male party members between eighteen and twenty-two join the armed forces. In addition, the party's youth organization, as well as the Society for Sport and Society, became very active in attempting to recruit young men for the KVP.

The KVP's officer corps was also still in the early stages of development. For example, in 1955, 74.5 percent of officers were below the age of thirty, and 73.3 percent came from either the agricultural or working-class segments of society. (In fact, one reason why the party had trouble recruiting officers was that it continued to show favoritism toward individuals from these classes.) Men from other sectors of society who might want to become officers had a much more difficult time proving their loyalty to the new regime. The decision to focus on individuals from the industrial and agricultural sectors was evident in soldiers' low educational level: 76.1 percent had only attended a *volkschule*, while 14.9 percent had graduated from a *mittleschule*. Only 9 percent had finished an *oberschule*. At the same time, party membership in the military had begun to climb. By this point 67 percent of all officers were party members.[52]

The NVA Makes Its Appearance

By the beginning of 1956, it was clear that West Germany would become a key part of NATO's military organization. From Moscow's standpoint, there was no reason to continue with the facade that the GDR did not have a military organization. Indeed, West Germany's decision to join NATO was like music to the East German political leadership's ears. After all, Walter Ulbricht had laid the groundwork in 1954 when he argued that if the Paris Treaty was ratified, the GDR would create its own military. Even though the actual sovereignty of East Germany's military would be limited because it was subordinate to the Warsaw Pact in a way that other East European states were not, the possession of an army would lend credence to the leadership's claim that the GDR was a sovereign state. The size of the new army—to be called the National People's Army or NVA—was set at 120,000. The 20,000-odd officers who had been serving in the KVP would be the core of the new military.[53]

On February 10, 1956, Willi Stoph, the GDR's first defense minister, signed the order for the creation of the NVA and the introduction of its new uniforms—which were reminiscent of those of those worn by the Wehrmacht in World War II. By April, the first unit—the First Mechanized Division—had become active. Additional units from company level to division and higher were added once they had been reformed and restructured to meet the demands of the Warsaw Pact.

In terms of the political situation, one of the first things the East German military leadership did was to create a new political officer school at Berlin-Treptow. The course of study was set at three years—a recognition that political officers had to be better trained if they were going to work effectively with line officers. By this point, political officers were being prepared for a variety of functions—because not all graduates of the school would immediately become deputies for political work. Some would become propagandists, while others would be given staff jobs where they were concerned with things like political training, work with the party's youth organization, or cultural work, to name only a few. The point was that these young men were being trained to serve in a wide variety of areas—a situation that left them with a basic, albeit very superficial, knowledge of many aspects of party work. They would have to learn how to carry out the job assigned to them once they arrived at their post. The same was true of the military aspect of their job. They would also have to learn their military skills on the job.

Not surprisingly, the Hungarian Uprising in 1956—in which the majority of soldiers in the Hungarian army sided with the insurgents—served as another wake-up call for the East German party leadership. The year 1953 had been bad enough, but what would happen now if something similar to the events in Budapest were to occur in East Berlin? Where would the army stand? Would it be loyal to the regime? In fact, the Central Committee meeting in July focused directly on the implication of the events in Hungary for the GDR. With an eye toward the possible internal unreliability of the NVA—which, after all, had been set up to protect the country from external threats—the party called for the strengthening of the so-called *Kampfgruppen*. These were paramilitary units set up primarily in factories. The party leadership believed that in a crunch, armed workers would be more useful in putting down popular uprisings than would the NVA.

As far as the NVA was concerned, the Ministry of Defense responded to events in Hungary by intensifying political work in the armed forces. In addition to political education for enlisted personnel, officers were pressed to increase their understanding of the intricacies of Marxism-Leninism. Ideological work would not be limited to formal courses, however. The party called for an intensification of work by agitators; more reliance on the party-controlled radio, newspapers, and journals; and other forms of ideological work. Political Activities rooms—where slogans, pictures, comparisons between East and West, and so on, were put up—also came to the fore. Anything to convince the soldier of just how lucky he was to live in a "Worker's and Peasant's State." And in an effort to drive home both the urgency and importance of intensified political work, the party made it clear that the commander (and by extension the political officer) bore primary responsibility for the unit's political and cultural work. If nothing else, the country's political leadership hoped this would be a way of ensuring that the need to tighten down ideologically would be taken seriously.

As we look at the role of the political officer at this time, it is clear that he would be held responsible for the actions of those who served in the unit—even if the commander was in charge. For example, Godau noted that if soldiers misbehaved—or God forbid, if one fled to the West—it was the political officer at whom a finger would be pointed. He obviously had not done enough to convince the individual concerned of the rightness and justness of the GDR, of how lucky that individual was to live in a state where exploitation by man of man was a thing of the past. It was the political officer's job to show soldiers that the GDR—not all of Germany—was their fatherland. The same was true of religion. It was up to the political officer to convince soldiers that going to church was not the thing that a good NVA soldier did. Besides, if the soldier ignored the political officer's advice and went to church anyway, this would inevitably be reported to the commander, who would then be criticized for failing to convince his troops not to attend church services.

Similarly, when a unit went on maneuvers, it was up to the political officer to explain to the troops why they were there and why they needed to perform this training mission in an outstanding manner—regardless of weather or other problems. Concerning the actual training exercise, the political officer stayed close to the commander to make certain that his orders were carried out in a timely fashion.[54] This was also true of discipline. If the unit had problems with alcoholism or with its training program, both the political officer and the commander would be called on the carpet.

Although we should not overemphasize the importance of party authority, we must not underestimate it. Let us assume, for example, that an officer wanted to marry a young woman who attended church. The political officer would try to convince the officer not to go through with the wedding, but if that failed, the only alternative would be the officer's dismissal from the service. The same was true of party membership for the officer. Almost 100 percent of all officers were members of the SED. If an individual was chastised by the party for an action—and that led to him being expelled from the party—he would shortly be ousted from the ranks of the NVA.

What about a situation in which the political officer decided to turn in his commander because of the latter's failure to show the requisite level of ideological commitment? (By this, I mean that the commander did not devote as much time to ideological indoctrination as the political officer believed appropriate.) According to Godau, that very seldom occurred. But when it did, he maintained that nothing happened to the commander.[55]

One activity that the army engaged in, which would play a major role in the collapse of the NVA more than thirty years later, was the practice of sending soldiers to help with the harvest. The country never had enough labor or machines when it came to harvesting, so the East Germans followed the Russian practice of using soldiers to help out—in exchange for a part of what was harvested going to the military. Political officers were often assigned to go with the soldiers. Their first task was to make certain that soldiers did only what was called for in the agreement the army had made with the agricultural enterprise.

From the standpoint of the political officer and the task at hand, this process would turn out to be a disaster. One thing that the party was striving for was to so isolate soldiers from the civilian populace that it could shape—if not dictate—their values and behavior. Over time, placing soldiers at the disposal of civilian firms (both industrial and agricultural) would undermine the political officer's attempt to inculcate a special weltanschauung among them. Too much contact with civilians was not good. More often than not, soldiers would bring negative civilian attitudes back with them to the barracks.

One thing was clear, however. Despite the party's and military leadership's efforts, the political situation within the armed forces was not improving. Something had to be done. This issue arose at a meeting of the Central Committee that was held at the end of January 1957. Discussions at this meeting emphasized that with the creation of the NVA, more would be expected not only when it came to military competency, but with regard to political reliability as well.

As a consequence, the Politburo issued an instruction (Instruction for Work by the Party Organization in the NVA) directed at making the party structure more effective when it came to raising the combat effectiveness of the armed forces. The primary purpose of this instruction was to provide more structure, both to control the activities in the armed forces and to enhance the party's presence in it. For example, in the latter instance, East Berlin decided that the Basic Organizations (*Grundoranisationen*) would be created any time at least three party members were present. In those cases where less than three members were present but candidate members were there, a Candidate Group was to be established. The idea was to make certain that party members were politically active at even the lowest level. They would not be able to escape the impact of politicization even if only a few of them were present.[56]

In addition to making the Basic Organization more effective, the new instruction put even more pressure on the political organs to administer and control all party activities within the armed forces. For example, the instruction clearly stated that all political work within the military would be led by the Political Administration. In essence, this meant that political officers and party organizations would get detailed help and supervision from higher-level political organs. Furthermore, a decision was also made to open up the position of party secretary (which existed at all levels) to individuals other than the deputy for political work—as had been the case in the past. Thus the party secretary—who would be responsible for party activities in the unit—could be another officer, an NCO, or, theoretically, even an enlisted man. The same was true of other unofficial party posts. The idea was to involve as many soldiers and officers as possible in party activities. Finally, the party made a conscious effort to select future political officers from the best and most politically active soldiers and NCOs. Once they had been chosen, they were sent to a three-year course, after which they would be assigned to a variety of political officer positions.

A year later, the Politburo issued another—and in this case, an even more important—document concerning political work. The 1957 instruction was a copy of one taken from the Soviet armed forces. In the eyes of the GDR's political and military leaders, the situation in their country was somewhat different and, as a consequence, required a slightly different kind of instruction. The party leadership had discovered that officers were not as dedicated to party work as it thought appropriate—especially in the aftermath of events in Hungary. Often they were too dogmatic and formal when it came to party work. Soldiers needed a clear statement of what the party's position was on key issues and that could only be accomplished if line officers took the role of ideology far more seriously.

Accordingly, the document placed particular stress on the close relationship between political and military factors. Everyone was supposed to be a committed and outstanding communist—he was only secondarily a soldier. In the USSR, the country had to wait more than ten years before the principle of unity of command had become common practice. In the GDR, on the other hand, it had been in operation since the KVP was created. Second, the instruction made it very clear that everyone—from the top to the bottom—was subject to party discipline. It did not matter what the individual's rank was: if he ran afoul of the party or failed to carry out an order issued by the party, the individual concerned would face disciplinary action. The same idea existed in the USSR, but it was seldom enforced. Criticism and self-criticism also were expected to become a part of everyday life. The East Germans believed that the existence of democracy in party structures was critical—that it would not work against military order and discipline, as some believed. In fact, a significant number of officers continued to try and dodge party work, believing that party democracy was incompatible with military order. "There are a group of comrade officers who have the false impression that the principle of unity of command, the right to give orders and the structure of the army are not compatible with the general principles of internal democracy."[57] But East Berlin's leaders were determined to implement what they understood to be internal democracy—at least, the kind that could be put into practice in the armed forces. For example, if the party put forth a candidate for party office and the majority of party members voted against him, he would not get the position. Such an approach would have been unthinkable in the Soviet military.

Third, the political administration of the armed forces was made directly subordinate to the Central Committee. "It must pay special attention to improving the content of political and ideological education of members of the military and their style of work."[58] Toward this end, greater emphasis was placed on ensuring effective party work in smaller and independent units—those often isolated from the NVA as a whole. Fourth, the role of the political organs in the development, advancement, and assignment of cadres was further strengthened. Indeed, when it came to personnel issues, the party organization was clearly in the driver's seat. The instruction also created Military Councils at the Bezirk, navy, and air force levels. These councils had the authority to issue orders—provided they were in accord with government and party instructions. Finally, the instruction called for ties between the military and local authorities to be strengthened. While the party wanted to control the environment in which soldiers lived to the maximum degree possible, it did not want an army that was isolated from the populace.

Former Members of the Wehrmacht

Former members of the Wehrmacht had continued to serve in the NVA, just as they had in the HVA and KVP. While they had been kept out of command positions, by and large, their behavior had been acceptable. However, the number of such individuals was gradually decreasing. In June 1951 there were 3,391 former Wehrmacht soldiers in the HVA. Of this number only 431 were officers, 959 were NCOs, and 2,004 were soldiers. By October 1953 only 4 percent of the officer corps had served in the Wehrmacht, but many occupied key positions. When the KVP became the NVA, at least four former Wehrmacht generals became NVA generals. However, by the middle of 1958 all four had retired. In fact, in 1956 there were only 500 former members of the Wehrmacht among some 17,500 officers.[59]

In February 1957 the party leadership decided to get rid of all those who had served in the Wehrmacht. Not only did such individuals have the wrong kind of class background, their presence in the NVA was being used by the West for propaganda reasons. Some specialists—those in the medical and educational fields—were allowed to stay on until about 1960, but the rest had to be out of the military by the end of 1959. By the middle of the 1960 there were only sixty-four former members of the Wehrmacht in the NVA.[60] Gradually, however, the issue went away and in 1971, any mention of prior service in the Wehrmacht was removed from the service records of those few who were still in the armed forces. Some of these men even managed to stay around until the NVA collapsed in 1989.

Ensuring Political Reliability

At the same time that former members of the Wehrmacht were being retired from the NVA, the party leadership introduced another policy—on a one-time basis. Following the Chinese example, the East German leadership decided that every officer should have had some experience in industry. A concern also arose that the gap between officers and enlisted personnel was getting too wide. As a consequence, the Central Committee decreed that all officers who had not worked in industry—and a large percentage of officers did not come from working-class families—should do so. As a result, in 1958 the party ordered all cadets and midshipmen who had not worked in industry to be assigned to those areas for specific periods of time. The next year the leadership ordered all officers to serve as simple soldiers for a certain amount of time. The purpose was to "raise their socialist consciousness" and to deepen their ties with the working class. The policy was ended in 1961 because all of the country's new officers arrived with a background in some part of the production process, as a consequence of the system of polytechnic education that had been introduced countrywide. What was interesting about this policy was that it was never utilized in the Russian military. While the NVA generally followed Moscow's line in most areas, in some cases it went its own way.[61]

Concerning the history of the NVA, the 1958 instructions would serve as the basic document governing the actions of political officers until the end of the GDR. Some modifications would be made here and there, but for practical purposes, the 1958 instruction would be the one that military and political authorities would refer to time after time, in explaining why party authorities in the armed forces acted in a certain way.

In terms of the political officer's training, another slight change was made at this point. On January 1, 1960, the political officer school's curriculum was modified. No longer would the school take young men from civilian life and train them to become political officers. No matter how hard they tried or what they did, these graduates were simply not on the same technical level as their line colleagues. As a consequence, the leadership decided to select political officers from among regular officers who had several years' service under their belt to attend the political officer school and, after completing it, be assigned as political officers.

The NVA as a "Coalition Army"

In the beginning of the 1960s, the USSR began to take a new look at the role played by East European militaries. With the adoption of what Moscow called the "coalition strategy," all of the countries of Eastern Europe were expected to assume a bigger role in the Warsaw Pact's military strategy. In essence, this new strategy called on the East Germans to play a key role in the pact's first echelon. Given East Berlin's desire to make the country indispensable to the Kremlin, the leadership wanted to do everything possible to assume such a role. The country, however, had a problem. If any more pressure was applied to the country's young men to join the military, many of them would simply have headed to the West. How to introduce conscription, a critical need if the NVA was to be expanded to the size required by the Soviets?

The answer came in the form of the Berlin Wall, which was constructed on August 13, 1961. While it would be an exaggeration to suggest that the wall was built solely to ensure sufficient conscripts for the NVA—the whole country was hemorrhaging people—it certainly helped solve the problem. At a minimum, the country's military leaders would have a predictable body of young men to draw on to staff the armed forces.

This decision meant a lot of work for political authorities within the military. They now had to spend a considerable amount of time explaining why the wall was built. In addition, as a sign of the increased importance the party was assigning to political work, the Political Administration was upgraded to a Main Political Administration (MPA). At the same time, the political sections of all major commands became political administrations.[62] Given the amount of work that would have to be done to justify the construction of a wall separating the two parts of Germany, as well as the need to improve military efficiency, it was not surprising that the political organs were expected to become more active.

The Volkskammer (the East German parliament) followed up on the building of the Berlin Wall by passing a law on conscription on January 24, 1962. Now, not only was the pool of recruits predictable, but all able-bodied young men would have no alternative but to serve in the NVA. Needless to say, the MPA recognized that the new situation would require modifications on how political work was conducted. Up to this point, the soldiers whom the MPA dealt with wanted to wear the NVA's uniform. Now, however, a lot of them had little interest in the military. And the problem was not just political. These young men would also have to be motivated to carry out combat training, to master technology, and to keep all of the equipment functioning.

Toward this end, party organizations, which had previously been present only down to the battalion level, were formed at the regimental and independent unit level. This meant that the basic organizations would not be enough—full-scale party organizations would now be created at those levels. Second, in order to ensure that East German officers were sufficiently trained to deal with this new brand of soldier, a system of social science training was introduced. In essence, this was an ideological training program that all officers were expected to complete. The program, which was part time, lasted four years. Once an officer finished, he would be given a diploma that was the equivalent of one given to a graduate of a party school. To make certain that every individual understood the seriousness of the situation, the Sixth Party Conference demanded that line officers perform better in the "single command" system by becoming more involved in political work.

The 1960s would represent the beginning of the professionalization of the NVA officer corps. East German forces were being asked to assume their place alongside the Soviet army in dealing with a perceived Western threat—an action that could only be achieved if the East German military was at its best. To quote Heinz Hoffmann, the long-time commander of the NVA,

> The main demand was to be constantly ready to stop a surprise attack working with the Soviet and other brotherly armies—under conventional conditions as well as by the use

of weapons of mass destruction. The structure of ground forces had to be modified so that they could be effectively protected from weapons of mass destruction.[63]

From a personnel standpoint, this mean that the old-line communists were no longer sufficient. They could not meet the demands of the new phase in East German military history. The individual officer and the NCO had to be first-class specialists as well as good communists. In time, this extreme pressure would reach the point where 85 percent of all personnel had to be present on their base at all time. In practice, this meant that all of the NVA's divisions (six active and five reserve) could be in the field and ready to go within two or three days. All navy ships had to be under way within sixty minutes, and half of them were on constant alert.[64] No other military in the world could have matched this level of readiness.

What was especially interesting in the 1960s and 1970s was the way in which the political organs began to emphasize the importance of technology. For those who have traditionally seen political organs in communist militaries primarily as vehicles to transmit the ideas of Marx and Engels, it must come as a shock to see political officers spending as much, if not more, time in their political lectures on the importance of mastering technology. The simple fact was that the military world was becoming increasingly complex, and this meant that both conscripts and professional soldiers had to take technology more seriously. The party went so far as to set up a secretariat within the Main Political Administration just to deal with technology, and a special Youth Commission was created to interest young people in technology.[65]

Political officer schools were again modified in 1963. In addition to proven line officers, soldiers who had served for at least six months and had done an outstanding job would be given the opportunity to attend the political officer school in Berlin-Treptow. Individuals whose parents had been members of the Nazi Party were prohibited from becoming political officers, as were candidates who had relatives of the first degree in West Germany. The minimum required education was a basic school completion and a blue collar job certification. In fact, a great amount of attention was devoted to finding candidates who had Arbiturs or who had completed at least the tenth class.[66]

One problem that plagued the NVA throughout its entire existence was the question of history. If the GDR was a unique and different state, then it must have a different kind of history. After all, it could not accept the idea that it was an artificial state created by the power of the Red Army. From a theoretical standpoint, it must be the culmination of a historical process. Accordingly, in July 1962, the party leadership called for more attention to be paid to the legacy of German history for the NVA. The result was the claim that the NVA was the first German worker and peasant state in the history of Germany. NVA soldiers and officers should understand the unique role they were playing. As a consequence, considerable effort was devoted to this topic, and several articles and books were written on it. The uprising against Napoleon or the uprisings of 1848–1849 were no longer just historical dates—they became important precursors in the development of the NVA. Likewise, the lives of German military figures such as Carl von Clausewitz and August Neihardt von Scharnhorst were reinterpreted, as the party decided that both men had been progressive reformers and therefore also important historical precursors for the modern leaders of the NVA. The reasoning was often a bit strained, but the important point was that the NVA now had a historical background, and political officers at all levels were now expected to learn history anew so they could pass it on to their fellow officers and men. The NVA represented the best in German history. This is how political officers were expected to explain why the NVA's uniforms looked so much like those of the Wehrmacht, while the Bundeswehr's were very different. The NVA looked German because it was German. The Bundeswehr's uniforms were an imitation of the American uniforms and therefore part of the imperialist camp.[67] At the same time, military installations, buildings, ships, academies, and so on were named for important historical figures. Furthermore, history was utilized by the NVA to try and build up support for close ties to the Russians. In October 1963, for example, a meeting was held between GDR and USSR historians on the 150th anniversary of the battle near Leipzig—with the clear goal of drumming up support for closer Russian-East German relations.

In addition to history, political officers would also have the task of explaining Moscow's invasion of Czechoslovakia in 1968. While some were able to do so, in other instances, certain political officers simply refused to go along. As a result of their outspoken opposition to Moscow's decision to destroy the Prague Spring, twelve political officers were tossed out of the SED and a short time later out of the NVA.[68]

The Life of a Political Officer

When it came to the political officer and his job, he remained under pressure to ensure the military and political readiness of his unit to the very end. There were some who would continue to show a lack of military knowledge; indeed, that was the main criticism raised against them by line officers.[69] Nevertheless, the political officer's main focus continued to be on topics such as discipline. If problems arose, he was expected to jump in and take care of them. How well were the soldiers doing their jobs? If something went wrong, he was expected to solve it. When, for example, an individual was successful in escaping to the West, the political officer would be closely questioned. Why didn't the soldiers perform as they were supposed to? Was it a result of a lack of ideological training? All negative information was supposed to be reported to the political officer. Take, for example, a case where a soldier was listening to Western radio programs. If the political officer learned of this, he would immediately call the soldier's platoon leader on the carpet to explain why he had not been doing the necessary political work. The political officer would then sit down with the soldiers to explain to them why listening to Western music was bad for them.

When it came to inspections by higher-level commands, the political officer was always directly involved. After all, he bore responsibility for the outcome. If troops were on the firing range, it was up to the political officer to motivate the soldiers. One political officer decided to make great use of a slogan the MPA had provided him—"Hit with the first shot." Or in another instance, the unit received new tanks. Obviously, the entire outfit had to be trained to deal with these new systems—and the political officer was expected to be involved. "They knew the real situation and could be found where the problems were the most difficult. Thanks to their work the division was able to accomplish the retraining successfully."[70]

Then there were morale issues. How to keep NCOs in the NVA when they worked incredibly long hours—fourteen hours a day, six days a week was normal! Part of the answer was convincing authorities to build NCO housing on or very near the base. A certain amount of "political watchdog" work was also necessary. Rothe tells the story of a political officer who came aboard a ship only to discover that the captain had a fur jacket that came from the West. When the political officer noticed the jacket, he immediately took the captain aside and told him to get rid of it!

The primary task of the political organs was to provide both guidance and support for political officers. When it came to the housing question noted earlier, the political officer—supported by his commander—would turn to these organs for assistance. If the political officer needed information on current party policy, it was up to the political organs to supply it. The same was true of newspapers, journals, or books. Did the political officers need new material to convince soldiers of the importance of technology? One couldn't expect an overworked political officer to come up with it. Were the party/political structures working effectively? If not, then why not? If there was a disciplinary problem—perhaps alcoholism or a wife beating—what was the political officer doing to deal with it? The political organs also had responsibility for military publications and for sports.

The political deputy was required to provide a monthly report, noting everything—positive and negative—that went on in his unit. This material was intended to help the next higher organ to evaluate the situation in the unit, and eventually, the report would work its way up the chain of command to the MPA. In addition, senior political officers would carry out periodic inspections. They would review the unit's activities, training programs, and personnel policies, and engage in conversations with a variety of soldiers and officers. The goal was to dig up problems—not only those that could impact on combat readiness, but those that might have a negative political effect as well.

The problem with this approach was that political organs often did not want to pass on negative information. If, for example, the head of the MPA were to give the defense minister a report that indicated problems in the NVA, the MPA head could expect to be called on the carpet. Why isn't the MPA doing more to deal with this problem? In the end, this situation would have disastrous consequences. It was one of the main reasons why the military collapsed. Although many of the men at the top were too doctrinaire (e.g., Defense Minister Heinz Kessler), most senior officers were caught unaware when mutinies and disciplinary problems became widespread.

The relationship between political officers and their commander was usually good. To be effective, they had to work together and, as a result, they seldom had any secrets from each other. Indeed, keeping secrets would have been a recipe for disaster for both of them. The rules for when and where political training was to be held were pretty much set in concrete—the equivalent of 16 hours per month for enlisted and 10–12 hours per month for officers. This situation would change only in 1989, as the military began to disintegrate and more time was devoted to lecturing the troops in an effort "to convince them that up was down."[71]

In an effort to improve the status of political officers, in 1971 the political officer's school was classified as a *hochschule*. This meant that the political officer would be entitled to receive a diploma upon graduation. His degree would be in social science. In addition to the greater prestige this provided, it also meant that when the officer left the military, he would have a degree to aid him in obtaining another job.[72] The education that political officers received changed again slightly in 1984. To begin with, for the first time, women were permitted to attend political officer school, and thirty-one of them became full-fledged political officers in 1988. Furthermore, given the technical demands that were being placed on the modern officer, 50 percent of their time was now devoted to military subjects, while the other half went to political topics.[73]

In 1975 a new policy was introduced—one called *Attestierung*. This was to become one of the most important personnel devices available to the NVA leadership. This policy required each officer to meet individually with his commander to discuss his current performance and plans for the future. Political officers were involved in this process to the degree that the party apparatus was part of any personnel decision, but (based on my discussions with a former senior NVA officer) it was his commander's evaluation that was most important. This evaluation covered not only his professional performance but his family life and other nonservice activities as well. For example, did the officer have children who were openly against the party? Did the officer's wife attend church? Was there a problem with alcohol? A second evaluation process was held in 1980, which covered the next five years.

The commander's statement would heavily influence what happened to him over the next five years. Was he to move up to a higher command position? Was he to attend one of the GDR's or USSR's higher military academies—crucial for advancing to senior positions? The only time political factors played an important role was if the officer was openly hostile to the party or had engaged in negative personnel actions.[74]

Regarding the actual job performed by a political officer, by the mid-1970s it had been fairly well standardized. To being with, he was responsible for the obvious—political education, a topic that went far beyond Marxism-Leninism. Its primary purpose was to motivate the soldier to do everything possible to raise his unit's combat readiness to the highest level possible. In fact, the topics the political officer covered in his lectures and discussions ran the gauntlet from the importance of proletarian internationalism (i.e., close ties with the Soviet Union) to the importance of cleaning one's rifle or writing to one's parents. The bottom line was that the political officer was expected to speak authoritatively on such topics. As one source noted, "They want (political) training that gives them something, but which also pushes them. They want a well-grounded, scientific, well substantiated answer to the pressing question; they want to learn and in the exchange of opinions measure their strength; not to be crammed full of 'gray' theories; they want the know the how of life, in order to carry out their military tasks."[75]

The political officer was also responsible for working with military families and for helping soldiers deal with their free time. If a family had difficulties in locating an apartment, the individual would go to the political officer for assistance. Similarly, if the soldier was having problems with his family, the political officer would be his first stop.

The political officer also spent considerable time on discipline and training. He worked on the training schedule, and he was the first one to deal with soldiers who had committed disciplinary infractions. He had to be sure that military and political matters were given equal time in training exercises. If too much stress was laid on one side, the other would suffer. Too much time spent on political lectures meant that the soldiers would not be able to carry out the military tasks assigned to them. At the same time, motivational lectures were crucial when it came to getting the individual to fulfill the most difficult tasks. Training played a central role in preparing the military for future combat tasks. It enabled the political structures to determine whether or not everything was coming together. Was morale at the right level? If not, then more political work was in order. The bottom line was that military exercises were about as close as one could get to combat.[76]

Although the control function of a political officer has often been exaggerated by Western analysts, there is no question that he played such a role, albeit to a very limited degree. When it came to overt signs of disloyalty, the security services (Stasi) had personnel assigned throughout the military to deal with such matters. A political officer was important because he knew the personnel in a unit, and if an individual began to show signs of disloyalty—or even signs of depression or fatigue—it was up to the political officer to make certain that this situation was brought to the attention of competent authorities.

The political officer was in charge of interfacing with outside organizations. Contact between civilians and members of the NVA was generally very restricted.

Often, military personnel would go for weeks, if not months, with no contact at all with civilians. However, someone had to carry out the liaison function because on occasions, joint celebrations were held or incidents occurred that required someone to step in and deal with civilian authorities. This was the political officer's job.

Competition (the so-called *Wettbewerbs*), in which various units vied with one another for prizes in military and political areas, also came under his purview. The idea was to use competition as a vehicle to improve the combat readiness of the NVA. Which unit had the most first-class technicians? The one that did would get a special reward. Or, which regiment scored highest on political tests or during training exercises? Someone had to be in charge and it was a political officer.

The company club—a place where soldiers could relax and enjoy themselves—was also the political officer's responsibility. NVA soldiers were permitted very little leave time, so a way had to be found to entertain them on base—this, too, was the political officer's task. Similarly, in an effort to raise interest in military technology, the political officer had the job of creating military circles where the latest equipment and weapons were discussed.[77]

One author summed up the characteristics such a political officer should have. He or she should be politically convinced of the rightness of the policies followed by the GDR, have the will and the ability to develop himself or herself politically, be in close contact with members of the military, have an exact knowledge of the interests and needs of the soldiers, be concerned about their service and living conditions, have the ability to emphasize the positive aspects of training, possess an unbreakable commitment to finishing something once it was started, and have the ability to win over the uncommitted to the party's goals.[78]

Conclusion

In 1990 the GDR and the NVA collapsed. The country became part of the Federal Republic, while most members of the NVA were dismissed, although a few were incorporated into the Bundeswehr.[79] Political officers were not permitted to join the Bundeswehr. As advocates for the despised world of communism, they were told that their services were no longer needed. Nevertheless, in retrospect, it is clear that they played an important role throughout the history of the NVA and its predecessors.

During the very earliest years of the HVA, political officers assumed a role similar to that played by commissars in the early days of the USSR or, to a lesser degree, during the French Revolution. They primarily had a control function because of the significant number of former Wehrmacht members who were being taken into the HVA and the many noncommunists who joined the police. Concerning their political or motivational work, it was very limited. After all, these were men who knew very little about Marxism-Leninism. They could read a lecture sent to them from East Berlin, but in fact they were ignorant on the topic. Political officers were also given a minimal education. They often found themselves working as political specialists after only a few weeks or months of political training. Militarily, they were often illiterate.

Soon, however, the situation began to change. By the time the KVP was created, political officers in the East German Army were playing a role very much like those in the Soviet military. They were no longer controllers. They were clearly subordinate to the commander—they were his deputy, not his equal. Indeed, this was the way things would remain until the collapse of the GDR.

As political officers, they played an increasingly important role in the military scheme of things. Indeed, one could argue that they were indispensable—given the nature of the communist system. They carried the ideological torch for the Communist Party. Equally important, however, was their pivotal military role. In one sense they were like the contemporary chaplain in the American military—concerned with issues such as morale, families, political socialization, and free time. In another sense, they carried out some of the tasks of an executive officer—worrying about discipline, training, housing, and meals. The same could be said of their work with entities outside of the NVA, where they took on the tasks of a public affairs officer. In time, they began to play an increasingly important role—one that was badly needed by the NVA leadership. This does not mean that incidents of conflict did not occur between line and political officers, but most of the East German officers I have spoken with did not remember the relationship as necessarily conflict-ridden. "They played their role, and I played mine," was the way one East German officer put it to me.

Admiral Hoffmann, the last commander of the NVA, probably said it best with regard to the importance of political officers when he observed,

> The majority of political workers in the army stood by their men, when the going got difficult, when there were difficult tasks to master, and they had to be concerned with problems of members of the army. The majority of them also had received a normal officer education and were concerned—whether as officers on duty or as a deputy to the commander—with working with their men in training or in combat service, and did not limit themselves to the direction of political training, cultural work, or work with the party and the FDJ basic organizations.[80]

Given how critical Hoffmann could be of other parts of the NVA when they failed to operate properly, this was a compliment.

The East German system failed not because the political officers did not do their job, but because the system was too dogmatic and conservative. The political structures carried out social analyses and polls. They knew what the problems were and understood what needed to be done to address them. Yet they were also part of the system, a system that was not able to adapt, a system that tried too hard to control everything from the top, a structure that was inflexible. As another veteran of the NVA put it,

For too long we had to come to terms not only with weaknesses and shortages, but to the contrary, also with this state's distortions and decadence and its obvious mistakes, even though they were at least partially visible prior to 1989 and had made us very angry.[81]

Political officers were part of the system and therefore part of the problem. However, they were also part of the reason why the NVA was so feared in the West and so respected by the Soviets and other members of the Warsaw Pact. They did their job well.

Notes

1. Stephan Fingerle, "Waffe in Arbeiterhand? Zur Rekruiterung der Offiziere der Nationalen Volksarmee," in Detlef Bald, Reinhard Brühl, and Andreas Prüfert, eds., *Nationale Volksarmee—Armee für den Frieden* (Baden-Baden: Nomos Verlagsgesellschaft, 1995), 119.

2. The word *kasernierte* is difficult to translate. Basically, it means police forces that are quartered in barracks, as opposed to police who go home every evening. The level of discipline and control in a *kasernierte* force is much higher than in a regular police force.

3. Rüdiger Wenzke, "Auf dem Wege zur Kaderarmee. Aspekte der Rektruitierung, Sozialstruktur und personellen Entwicklung des entstehenden Militärs in der SBZ/DDR bis 1952/53," in Bruno Thoss, ed., *Volksarmee schaffen—ohne Geschrei* (München: R. Oldenbourg Verlag, 1994), 223.

4. Wenzke, in Thoss, *Volksarmee schaffen,* 221.

5. Wenzke, in Thoss, *Volksarmee schaffen,* 225.

6. Wenzke, in Thoss, *Volksarmee schaffen,* 256.

7. As cited in Wolfgang Eisert, "Zu den Anfängen der Sicherheits-und Militärpolitik der SED-Führung 1948 bis 1952," Thoss, *Volksarmee schaffen,* 171.

8. Eisert, in Thoss, *Volksarmee schaffen,* 187.

9. Hoffmann refers to Heinz Hoffmann, who played a major role in the creation of the East German police/military forces from the earliest period up to 1982.

10. Eisert, in Thoss, *Volksarmee schaffen,* 187.

11. Werner Rothe, *Jahre im Frieden: Eine DDR-Biographie* (Berlin: GNN Gesellschaft für Nachrichtenerfassung und Nachrichtenverbreitung, 1997), 17.

12. Peter Jungermann, *Die Wehrideologie der SED und das Leitbild der Nationalen Volksarmee vom sozialistischen deutschen Soldate* (Stuttgart: Seewald Verlag, 1973), 85.

13. Wenzke, in Thoss, *Volksarmee schaffen,* 232.

14. Kurt Schützle, "Zur geschichtlichen Entwicklung der Hauptverwaltung für Ausbildung (HVA) 1949–1952," unpublished manuscript, 4. This manuscript was previously classified by East German authorities, presumably because it dealt in detail with the creation and build-up of the HVA, including Soviet involvement, something that the GDR officially denied took place throughout its existence.

15. Heinz Godau, *Verführter Verführer, Ich war Politoffizier der NVA* (Köln: Markus Verlag, 1965), 22.

16. Wenzke, in Thoss, *Volksarmee schaffen,* 243.

17. Schützle, "Zur geschichtlichen Entwicklung der Kasernierten Volkspolizei," 8.

18. Schützle, "Zur geschichtlichen Entwicklung der Kasernierten Volkspolizei," 17.

19. Autorenkollektiv des Deustschen Instituts für Militärgeschichte, *Zeittafel zur Militärgeschichte der Deutschen Demokratischen Republik 1949 bis 1968* (Berlin: Deutscher Militärverlag, 1969), 19.

20. Joachim Schunke, "Von der HVA über die KVP zur NVA," in Wolfgang Wünsche, *Rührt euch! Zur Geschichte der NVA* (Berlin: Edition Ost, 1998), 41.

21. Eisert in Thoss, *Volksarmee schaffen*, 197.

22. Wenzke in Thoss, *Volksarmee schaffen*, 256.

23. Fritz Elchlepp, Tape 22, in Joachim Eike, *Die verschwundene Armee*, interviews, which were prepared for a program of the same name and provided to the author by Herr Eike. Cited hereafter as Eike.

24. Schützle, "Zur geschichtlichen Entwicklung der Kasernierten Volkspolizei," 13.

25. Fritz Streletz, Tape 10, in Eike.

26. Schunke, in *Wünsche*, 45.

27. Schunke, in *Wünsche*, 45.

28. Schützle, "Zur geschichtlichen Entwicklung der Kasernierten Volkspolizei," 21.

29. Erich Hagermann, *Soldat der DDR* (Berlin: Verlag am Park, 1997), 78.

30. Rothe, *Jahre im Frieden*, 34.

31. Wenzke, in Thoss, *Volksarmee schaffen*, 246.

32. Godau, *Verführer, Verführrter*, 26.

33. Hagermann, *Soldat der DDR*, 104–105.

34. Herbert Peter, "Als Politoffizier in den Vorläufern der NVA," *Militärgeschichte* 6 (1990): 583.

35. Schunke, in *Wünsche*, 44.

36. Godau, *Verführer, Verführrter*, 46, 52

37. Rothe, *Jahre im Frieden*, 34.

38. Hagermann, *Militärgeschichte* 7 (1997): 11.

39. Schützle, "Zur geschichtlichen Entwicklung der Kaser nierten Volkspolizei," 31.

40. Schunke, in *Wünsche*, 58. Throughout this chapter the traditional German ß is being used for the letters ss.

41. Hagermann, *Militärgeschichte*, 127.

42. Hagermann, *Militärgeschichte* 7 (1979): 11.

43. Hagermann, *Militärgeschichte*, 12.

44. Hagermann, *Militärgeschichte* 7 (1979), 10.

45. Godau, *Verführer, Verführrter*, 73–74.

46. Wenzke in *Thoss*, 267.

47. K. Baarss, in Eike, Tape 25.

48. Godau, *Verführer, Verführrter*, 95.

49. Godau, *Verführer, Verführrter*, 99.

50. Wenzke, in *Thoss*, 266.

51. Schützle, "Zur geschichtlichen Entwicklung der Haupverwaltung für Ausbildung," 21.

52 . Schunke, in *Wünsche*, 66.

53. Wenzke, in *Thoss*, 271–272.

54. Godau, *Verführer, Verführrter*, 153.

55. Godau, *Verführer, Verführrter*, 134.

56. Werner Hübner, "Zur Rolle der Partei in der Nationalen Volksarmee," in *Wünsche*, 425.

57. Hübner, "Zur Rolle der Partei," in *Wünsche*, 422.

58. Hübner, "Zur Rolle der Partei," in *Wünsche*, 419.

59. Reinhard Brühl, "Zur Militärpolitik der SED-Zwischen Friedens Ideal und Kriegsapologie," in Detlaf Bald, ed., *Die Nationale Volksarmee* (Baden-Baden: Nomos, 1972), 144–145.

60. Dr. Wenzke, Tape 23, in Eike.

61. Brühl, "Zur Militärpolitik der SED," 123.

62. G. Lux, "Die Entwicklung des Aufbaus der Politorgane und Parteiorganisationen der SED in der NVA," *Militärwesen* 6 (1974): 78.

63. Heinz Hoffmann, *Sozialistische Landes Verteidigung, Aus Reden und Aufsätzen, 1974–1978* (Berlin: Militärverlag der Deutschen Demokratischen Republik, 1979), 240.

64. "Wir müßen unsere Feldwebel neu backen," *Die Welt*, 1 December 1990, and Theodor Hoffmann, *Kommando Ostsee* (Berlin: Verlag E. S. Mittler, 1995), 212–213.

65. Lux, "Die Entwicklung des Aufbaus der Politorgane," 78.

66. Heinrich v. zur Mühlen, "Der Rekrut und der Kommissar," *SBZ-Archiv* 14, no. 3: 40.

67. Paul A. Koszuszek, *Militärische Traditionispflege in der Nationalen Volksarmee der DDR* (Frankfurt am Main: Hagg & Herchen, 1991), 138.

68. Wenzke, Tape 23 in Eike.

69. Rothe, *Jahre im Frieden*, 84.

70. Rothe, *Jahre im Frieden*, 99.

71. Eberhard Haueis, "Die Führende Rolle der SED in der Nationalen Volksarmee," in *Wünsche*, 441–442.

72. Autorenkollektiv, *Zeittafel zur Militärgeschichte der Deutschen Demokratischen Republik 1969 bis 1977* (Berlin: Militärverlag der Deutschen Demokratischen Republik, 1979), 36.

73. Kurt Held, Heinz Friedrich, and Dagmar Pietsch, "Politische Bildung und Erziehung in der NVA," in Backerra, *NVA: Ein Rückblick für die Zukunft*, 214.

74. Based on a conversation with Colonel Hans-Werner Weber, who spent thirty-five years in the NVA and its predecessor.

75. K. Freudenreich, "Zu einigen Kennzeichen der Schulungsgruppenleiter und Teilnehmer an der politischen Schulung," *Militärwesen* 10 (1975): 32.

76. S. Schatz, "Aspekte und Erfahrungen der Führung der politischen Arbeit bei Truppenübungen," *Militärwesen* 1 (1980): 22.

77. See Hans-Joachim Belde, "Der Politstellvertreter in der NVA—Multiplikator im Ausbildungsdiest oder Politruk," *Wehrwissenschaftliche Rundschau* 6 (1979): 181.

78. K.-H. Licht, F. Simmerl, and P. Stahlberg, "Persönlichkeitstheoretische Aspekte der Erziehung und Ausbildung des Politstellvertreters," *Militärwesen* 4 (1983): 34.

79. For a detailed discussion of this process, see my *Requiem for an Army: The Demise of the East German Military* (Lanham, Md.: Rowman & Littlefield, 1998).

80. Hoffmann, *Kommando Ostsee*, 78.

81. Kurt Held, "Soldat des Volkes? Über das politische Selbstverständnis des Soldaten der Nationalen Volksarmee," in Bald, *Die Nationale Volksarmee*, 47.

Conclusion

Nothing endures but change.

—Heraclitus

WE ARE NOW AT THE POINT where it is time to answer the question posed at the beginning of this study. To what degree are American chaplains comparable to similar positions in Cromwell's army, to the French commissars and commissars in the Soviet Revolution; to the NSFOs; and to the Soviet and the East German political officers? Did all of these positions share a core set of functions? Is it fair to say, as a Soviet political officer in Vladivostok said to me in 1989, "Political officers are just like chaplains. After all, we do the same thing!"

Similarly, it is also time to deal with the issue of change in totalitarian regimes. Is it possible to relate changes in the nature of the tasks carried out by political commissars and political officers with the nature of civil-military relations in such political systems? If it is, this will suggest that a greater possibility to detect change exists in such systems than some believed to be the case during the days of the Cold War. Nothing in politics stays the same, and that includes civil-military relations. The task is to detect change in systems that are highly secretive and closed.

Before discussing these issues in any depth, however, let us turn to Table 2. If this table demonstrates anything, it is that all of the institutions included in this study carried out certain common functions—in spite of their differences. Indeed, I would argue that as hypothesized in this book's introduction, all of the officers discussed were involved in the core functions of motivation, building morale, and political socialization. The chaplains in the American military and in Cromwell's army may have been primarily concerned with spiritual matters, while

TABLE 2
Functions of Chaplains, Political Commissars, and Political Officers

	Provide Spiritual Comfort	Motivate, Counsel Soldiers	Political Socialization	Change Basic Value System	Control Actions of Line Officers	Assume Command
Cromwell's Chaplains	X	X	X			
French Revolution	X	X		X	X	
Soviet Commissars	X	X		X	X	X
German NFSOs	X	X				
Soviet Political Officers	X	X				
East German Political Officers	X	X	X			
American Chaplains	X	X	X			

Soviet and East German political officers focused on Marxist ideology—but they all had the tasks of keeping up morale, motivating soldiers, and socializing them.

The table also shows that only two—and perhaps three—of these institutions were ever engaged in a control function. The French representatives on mission, as well as the Russian commissars, exercised control over line officers, as did the East German political-cultural officers during their very short existence. The German NSFO was moving in that direction and could have become a control mechanism, but the country and its military collapsed at the end of World War II.

Common Functions

It made little difference what the ideological bias of the regime was, although its ideological orientation had an impact on how these problems were dealt with and where emphasis was placed. Chaplains led religious ceremonies, while political officers and political commissars gave ideological lectures. All of the officers discussed in the pages of this book spent a considerable amount of their time working in areas such as morale, motivation, and political socialization.

Morale

The question of morale was omnipresent, regardless of the time and place or the military concerned. It mattered little whether the situation was Cromwell's troops fighting the king, Soviets battling the Germans in World War II, or Americans trying to control Vietnam. Without solid morale, an army's chances for success on the battlefield are very limited. Indeed, I suspect that if one asked a cross-section of officers what is the most important factor in combat, morale would rank very high, if not at the top. The problem, however, is how to achieve it. A commander could devote most of his time to the issue and would probably succeed in raising morale. General George Patton and Soviet Marshal Georgi Zhukov were certainly very effective in this regard. However, both of them relied heavily on their chaplains and political officers to take the kinds of actions that kept morale at a high point. Both men were fully aware that they could not devote most of their day to such endeavors. It would have meant taking crucial time away from issues like training, weapons, logistics, equipment, war planning, and a hundred other things that kept commanders busy. The amount of time required to counsel soldiers, to try and solve their many personal problems, and to provide lectures and talks to socialize them to the military's goals and raison d'être went beyond the abilities of even the most dedicated and hard-working commander. Patton and Zhukov might have been responsible for the final product, but expecting either man to carry out all of these tasks was not realistic.

What to do? One of the things I found most interesting in looking at these various militaries was that they all came up with a military specialist to carry out this task. American chaplains worried about race relations just as Soviet political officers did. Both were faced with the need to convince the majority population of the importance of respecting the cultural and historical achievements of minorities, regardless of whether they were Central Asians or African Americans. Both chaplains and political officers became involved in affirmative action programs. The commander might preside over ceremonial actions, but it was the chaplain or the political officer who did the hard work of trying to convince soldiers of the need to get along with individuals from a variety of racial and ethnic backgrounds.

Motivation

Both chaplains and political officers were expected to play a key role in motivating the troops. They were there to convince soldiers of the need to improve themselves, to become better qualified at carrying out their military duties. This was one of the main reasons a lot of line officers accepted the presence of chaplains in the American military. Leaving aside the religious function, many line officers believed that whatever chaplains' shortcomings (e.g., working to end the brutal discipline that existed on ships in the nineteenth century or carrying on antidrinking campaigns at army bases), chaplains helped to motivate members of the military to become better soldiers.

Discipline was also important. While American chaplains were not always directly involved in the disciplinary process (although, at one time they were expected to serve as defense counsels for those accused of wrongdoing), they played an important, indirect role by counseling soldiers and by trying to help them solve personal problems that got in the way of their ability to perform their duties. This was one task assigned to Cromwell's chaplains—help maintain good order and discipline, while also working closely with the civilian population. Although times have changed, the tasks facing American chaplains in the past fifty years have been similar—lectures on AIDs, drugs, and relations with the local population.

Political officers also spent a considerable amount of time working on such issues. In fact, in many cases, they were directly responsible for discipline. If discipline deteriorated, then the political officer—regardless of whether he was an East German or a Soviet—was not doing his job and his career could suffer. The same was true with regard to training. Political officers hammered soldiers on the importance of keeping up with the latest weapons technology—of the value of winning the next highest medal for technical achievement. And like chaplains, they were expected to be present to help motivate and inspire soldiers to carry out their tasks more effectively and efficiently when they went to the field. When an inspection team from Berlin or Moscow visited the troops, one of the first things its members did was to talk to the unit's political officer about the unit's technical qualifications. If these were low, the political officer had better have a plan in place for raising standards, if he hoped to keep his job.

When it came to combat, chaplains and political officers were crucial. On the night before a battle, Cromwell's chaplains worked to inspire soldiers by holding prayer and religious singing services. Or, they might join troops in fasting in order to better focus their attention on the upcoming conflict. Likewise, when battle was joined, the presence of these clergymen at the front lines did much to inspire the troops. The same was true of chaplains in the American military over the years. One can only imagine the impact that Chaplain O'Callahan's actions on the USS *Franklin* had in motivating others to perform at a level many would have believed impossible. His role in throwing ammunition over the side, even though it was burning his hands, or his forays inside the smoke-filled, burning ship helped convince others that they could and should give everything they had to save the USS *Franklin*. If chaplains like Father O'Callahan could show such incredible bravery and disregard for personal safety, others could as well. In World War II, even the Germans recognized that they could not do without chaplains. They may have found chaplains' message repugnant, but the Nazis knew full well that chaplains played a key role in helping to give the soldiers the will to go on and fight what was clearly a hopeless battle.

The heroism of commissars during the French and Russian Revolutions should not be discounted as a motivating force. There are too many stories about commissars who, in pursuit of the communist dream, sacrificed their lives. Their willingness to lead their men into battle during the French Revolution was critical. Similarly, the ability of Soviet commissars to withstand the greatest rigors helped win the war against the Whites. And the same thing could be said about Soviet political officers during World War II. Their casualty rates were among the highest in the Red Army. Not all were heroes, to be sure—any more than all chaplains fit the mold of Chaplain O'Callahan—but those who were, made an important contribution toward victory in both armies.

In fact, the importance of both the American chaplain and the Soviet political officer serving as role models in World War II should not be underestimated. It was not an accident that Berlin kept complaining about the quality of individuals sent to become NSFOs. Senior party officials knew that the kind of officer who would get the highest respect and do the most to motivate soldiers was an individual who had seen combat firsthand and who had shown considerable bravery in the process. One could argue that in addition to opposition from line officers who were jealous of NSFOs' command prerogative, the main reason NSFOs exerted very little influence was because they commanded very little respect from the average soldier. A school teacher? A former theology student? An officer who had spent most of his career in staff jobs? Be serious! What did such individuals know about the horrors and rigors of combat? One can almost hear the troops shouting, Send us real soldiers!

Political Socialization

Although differences in form and substance existed in the different militaries, all chaplains and political officers had one primary goal: convince the soldier or sailor of the rightness of the cause he was fighting for, and how disastrous the situation would be if his country did not win the war. Not all chaplains and political officers were expected to change the soldiers' value systems. For example, this was not an issue for Cromwell's chaplains. Rather, for them it was a matter of seizing upon the existing values as a means to help explain the religious significance of the struggle against Charles I and the Anglican Church to the troops. The message was rather simple: If Cromwell's men did not prevail, their ability to practice their religion as they saw fit would be eliminated.

For Soviet political officers during World War II, it was a combination of explaining the horrendous nature of German atrocities while at the same time emphasizing the positive role the Communist Party was playing in driving out the Germans. War on the Eastern Front was different from that in the rest of Europe—it was a struggle for survival. And it was up to the political officers to make that crystal clear to their troops.

After the war, when the United States became Moscow's primary enemy during the Cold War, the political officer's job became one of explaining why the Soviet cause was just and American or Western plans were evil. This was not an easy task, given the disparities in living standards not only between East and West, but between the USSR and Eastern Europe. The Germans lost the war. Yet East Germans lived far better than citizens of the USSR. Why? The function had to be performed. How could the Kremlin expect to triumph if Soviet soldiers were not convinced that their cause was just? Again, it was up to political officers to make the Kremlin's case to the troops in a convincing manner.

In a certain sense, when it came to political socialization, the East German po-
litical officer's task was even more formidable than that of his Soviet counterpart.
Most East Germans had little love for the Russians, and the majority looked upon
communism as a foreign import. Indeed, most of them would have agreed with
Stalin's purported answer to a question about the possibility of communism in
Germany—that communism fits Germany the way a saddle fits a cow! The East
German political officers had to be very creative—they had to try and convince
their own countrymen that the Soviets were their friends. This in the aftermath of
Soviet atrocities in Germany and Berlin at the end of the war. In addition, political officers had to convince their troops that the obvious difference in living standards between East and West Germany was a result of the "superiority" of the East
German system. One can imagine just how hard this must have been, considering
that the living standard in the West was two to three times higher than in the East.
The situation in East Germany was so artificial that East German political officers
were forced to help "invent" a new historical past—by convincing their troops that
just as the GDR was fundamentally different from the FRG, the NVA was a totally
new kind of army and bore no resemblance to the Bundeswehr—not an easy task.
The NVA was not only different but it came from a "Worker's and Peasant's State,"
and this made it morally, politically, and ideologically superior. In essence, it was
up to political officers to convince their troops that what the rest of the world believed to be up was really down and vice versa.

American chaplains have not traditionally seen themselves involved in the
process of political socialization. Indeed, many would argue that was not their
"job." Furthermore, from a bureaucratic standpoint, they have often shied away
from such a task for the simple reason that many believed that it would give ammunition to those who wanted to do away with chaplains because they supposedly violated the Establishment Clause of the First Amendment. In spite of that
hesitation, however, chaplains have played an important role in this area. At pivotal points in American history they were in charge of patriotic education of the
troops. And for anyone to suggest that chaplains have not made a close connection between the American system and religion during periods of combat, the argument would seem far-fetched at best. How else would a chaplain answer the
question "Why are we fighting, Padre?" Most probably, by stating that "We are
fighting for freedom, democracy, our families, and the right to worship God as we
see fit. If we fail, the Germans or the Russians or whoever will be running our
country and all of our freedoms—including our right to practice our religion—
will be taken away."

The French and Russian Revolutions were different when it came to the process
of political socialization. In these cases, political commissars could not rely on
preexisting value systems to support their arguments. In both instances it was a
case of bringing about major changes in the majority of soldiers' thinking. Imagine a French or Russian peasant being offered a rifle and being told that he was
fighting for the "working class" or for liberty and equality. Both concepts would

have been very foreign to him. The task for the commissar was to eradicate the old ideas and replace them with new ones—not an easy undertaking under the best of circumstances. All kinds of props—many rather primitive at the time—were utilized. Songs, pamphlets, lectures, newspapers, and whatever seemed to be most useful were seized upon to convince an often skeptical army of peasants of the advantages of the new system vis-à-vis the old one. The task was made even more difficult by the fact that most Russian and French recruits for the revolutionary armies were illiterate.

The German case is again somewhat different. There really wasn't any need to convince the majority of soldiers of the legitimacy of the Third Reich. They and their officers accepted Hitler and the Nazi Party. Besides, there were also chaplains in the Wehrmacht. While one suspects that few of them would have drawn a direct line between God's will and the Third Reich, chaplains do appear to have played a positive role, in that they satisfied the spiritual needs of the soldiers concerned and in the process made them into better soldiers.

The role of the NSFOs in political socialization was directed primarily at convincing German soldiers to fight fanatically for a dream and an empire that was collapsing all around them. To make German soldiers believe that it was really a battle between the forces of good and evil—Judaism and Bolshevism versus German civilization—was no easy task. However, the message was as clear as it was simple—the two were incompatible and only one would survive. Either Germany won the war in the East, or German women and children would become the slaves of the worldwide Bolshevik-Jewish conspiracy.

Let us now turn to the issue of change in totalitarian systems.

Detecting Change in Totalitarian Systems

One conclusion that jumps out at the analyst is that the presence of political commissars undercut military efficiency. This was even more true of representatives on mission. There were a number of reasons for this situation.

Clear Lines of Authority

A system of dual command (as the Soviets and East Germans referred to it), whereby both the commander and the commissar had to sign an order for it to be valid, created confusion in the minds of soldiers. Whose order should I obey? What if the commander tells me one thing, but the political commissar says the opposite? Such a situation could be catastrophic on a battlefield. While the need for clear and unambiguous lines of command is obvious at present when we are faced with "push button" wars, it was just as real during the French or Soviet Revolutions. Armies are only effective if they work together—in unison. Under normal circumstances, they cannot both retreat and advance at the same time. They must be given clear and unambiguous orders if political and military authorities hope to make effective use of them and avoid defeat. In many cases, armies wait for orders and if these are not forthcoming in a clear and unambiguous fashion, the result could be disastrous.

The confusion that the presence of political commissars created within an army went all the way to the top. It was not just a concern of the average soldier in a fox-hole. Let us assume that we were dealing with a regimental level (or higher) unit. What if the commander and the political commissar disagreed on how to plan operations? In the French Revolution, this would not have been a problem (at least, under the Jacobins) because the orders of the representative on mission overrode those of any commander. In fact, the representative on mission could appoint new commanders on the spot. Political commissars could make life difficult for a commander during the French Revolution by submitting a negative report—which might lead to the commander's recall—but political commissars do not appear to have enjoyed the right to countersign orders. They were also intimately involved in the legal system, but here, too, their power was limited. Nevertheless, from the standpoint of military efficiency they played the role of a nuisance factor—commanders feared them, and the average soldier never knew where the commissar stood in the chain of command. The chain of command was confused at best.

But what about the Russian Revolution? Under the rules of the dual command system, both the commander and the commissar had to sign an order—so, in a certain sense, both were equal. Equality, however, has always been foreign to effective military organizations. This was especially true in the Russian Revolution, when the very existence of the new political system was at stake—a failure to act could be just as decisive as a bold action. In this sense, the lines of authority were even more confused than they had been in the French Revolution. In the latter instance, it was clear that representatives on mission had unlimited power, whereas in the Russian case the dilution of power between the commander and the commissar could easily lead to confusion. Lenin was smart enough to convince his colleagues to make some concessions by relying on former members of the Imperial Army, but the concern for political reliability necessitated the presence of commissars. The same situation existed in East Germany in the beginning when the country's paramilitary forces were so dependent on those who had served in the Wehrmacht during World War II. A way had to be found to ensure their loyalty. The answer was the short-lived experiment with political-cultural officers.

Amateurs or Experts?

Another problem with the commissar arrangement—and this was particularly true in the French case—was that it meant putting amateurs in the position of senior military commanders. There were occasions when such an arrangement was effective—the work of Jeanbon Saint-André in rebuilding the French fleet in Brest is a case in point. In many other instances, however, interference by civilians made the situation worse. The actions by the Jacobin Clubs did more to undermine military efficiency than anything the monarchists could have dreamed up. Instead of reinforcing discipline, by questioning the qualifications and fitness of the army's officers these civilian clubs undercut it constantly.

The situation was less of a problem in Russia, simply because civilian organizations such as the Jacobin Clubs did not exist. But here, too, problems could develop—if for no other reason than that a strong difference of opinion between the military specialist and the commissar could lead to gridlock. This was an especially dangerous situation (from a military standpoint) if the disagreement was caused by ideological differences. By that, I mean a situation in which a political commissar tried to impose an "ideologically acceptable" solution on a line officer—even if it made no sense. For example, let us assume that the situation called for the unit to go on the defensive until reinforcements arrived. While that might make military sense, a political commissar, under the influence of Bolshevik propaganda, could insist that the only correct approach was to go on the offensive. Yet from a purely military standpoint, the lack of proper weapons, the wrong weather, or a shortage of troops could lead to disaster, regardless of ideological purity. Which approach to adopt? One was ideologically sound; the other made military sense. While the two officers tried to sort this matter out, the military forces could have faced a disaster. And even if it turned out that the line officer was right and won the day, if he followed an ideologically unsound strategy, he ran the risk of being called to task at a later date.

Undercutting Initiative and Creativity

Another negative effect of the commissar arrangement was that it undercut initiative and creativity on the part of military officers. Individuals in senior positions could not possibly script every possible action—even though they often tried to do just that. Once the battle started, armies were forced to rely heavily on the commander's willingness to take chances, to take the initiative. However, the commissar system was all about control—control over every part of military life. What intelligent military officer would take the initiative in the field if he was concerned that his every action was being watched by someone above him—especially if that person was not a military expert? It was much easier (and safer) to sit back and be reactive. This is exactly what happened in many instances where commissars were present. Indeed, one could argue that a legacy of the commissar system in the Soviet and even the Russian Army was that it helped create a mindset that was—and still is—very suspicious of initiative and creativity. Staff officers in the Russian Army continue to try and micromanage every action on the battlefield, leaving little room for initiative and creativity.

Why the Commissar System?

Given all of these negative aspects of the commissar system, why institute it? The answer was simple. The new French, Russian, and later the German elites in World War II had no alternative. They were engaged in a life and death struggle. They needed every soldier they could get their hands on. This was especially true of those who had served in the Imperial armies, the so-called military specialists. Even two hundred years ago, military expertise was not something that was transferable between civilian and military worlds. For example, artillery was a complex area. One could not just put a civilian in uniform and expect him to be able to perform the complicated actions of an artillery officer without special training and education. Figuring out the proper elevation of guns and the impact of weather and winds, not to mention the correct munitions, required many hours of specialized training.

The situation was not much different when it came to the more simplified worlds of the infantry or cavalry. Moving large numbers of troops around the countryside was a lot more complex than many civilians realized. What if a unit were transferred from one place to another and the needed supplies were not on hand? What about maneuvering troops in combat? Keep in mind, the other side was staffed almost entirely by professional soldiers who were very experienced in this area. The bottom line was that without these military specialists, the war would be lost and with it the new communist political system.

Faced with such a situation, the revolutionaries made a compromise—they decided to put up with a less-than-perfect situation. And here, it is important to note that the situation in France was different from that in Russia. The former revolution (especially under the Jacobins) was more radical than the Russian version. As a consequence, while the French relied on military officers from the old regime, they did not give the officers as much decision-making latitude as did the Russians. This was especially true of the period after General Charles Dumouriez defected to the monarchists. The primary lesson that most revolutionaries took from that act of "treachery" was that few military officers from the former imperial military could be trusted, no matter how closely they were watched.

Although some problems arose, the Russian approach to utilizing political commissars was more systematic than in France. The relationship between the two individuals was clearer from the start. Both officers knew where they stood. In many cases, this arrangement had the positive impact of forcing both officers to cooperate. They were in the same boat. Unless a political commissar could prove that the line officer or commander was clearly out of line, failure on the part of the unit would be laid at both men's doors.

In the East German case, the problem was not a war; it was building an army while faced with an open border that enabled soldiers to move back and forth. Political-cultural officers were a short-term solution. Something had to be done to ensure that those who were recruited into the paramilitary forces would behave.

Abandoning Political Commissars

Once the emergency had passed, a new approach was needed. The commissar system made sense as long as the system had to rely on potentially unreliable officers from the old army, but now it was time to come up with a more efficient structure. This was done by gradually getting rid of those military specialists who were neutral or even hostile to the new regime. Indeed, if officers could not convince authorities that they were fully devoted to the goals of the revolution, they would be let go. At the same time, a cadre of military specialists had to be kept around. After all, they were needed to train the younger generation of ideologically committed but militarily illiterate men and boys who were now joining the military. In time, however, this cadre would also become expendable and, for the most part, would pass from the scene.

At the same time this process was going on, important structural modifications were underway. In this case, the Russian experience was most illustrative. Those officers who had proven their loyalty to the system were gradually given more authority. Over time, the dual command system was transformed into the single command system. Indeed, by the end of the 1920s, in the vast majority of cases commanders had complete authority over all activities in the units they commanded.

The Russian or Soviet case also illustrates how flexible the institution of political commissars could be. When Russia was faced with a threat, commissars could be reintroduced, even if in a modified form. This is exactly what happened in the USSR in 1937, when the commander's authority over political matters was returned to political commissars—as a result of the chaos created by Stalin's purges. Equally important, however, was the refusal of the Kremlin to grant these political commissars the kind of dual authority their predecessors had enjoyed during the civil war. Value congruence was not a serious problem. Rather, the issue was the absence of large numbers of line officers (who had been purged) and uncertainty over the correct political line. The purges created major concerns over potential political instability within the army—at least, in Stalin's mind. Something had to be done to stabilize the armed forces. Once the situation had become more stable, the military reverted to reliance on political officers.

Then came the German onslaught against the USSR. Thousands of reservists were called to the colors. The Kremlin was faced with one of the most massive training programs imaginable. To deal with the situation, in July 1941 Moscow again introduced a modified form of political commissars. Once again, the issue was not a lack of value congruence. Instead, there were not enough commanders able to deal with matters in both the political and military spheres, so the work was again split up—commanders would worry about military matters, while political commissars dealt with political questions. Each was supreme in his own area. In the fall of 1942, as line officers became more proficient in political areas, the commissars were again abolished in favor of the political officer system. The commander was in charge and the political officer became one of his deputies—an arrangement that would continue until the Soviet Union broke up many years later.

The German Experiment

Although the Germans did not get around to creating a full-blown political commissar system until the very end of the Second World War, the battle over the National Socialist Leadership Officer demonstrated that there was open opposition on the part of the regular military to the appearance of anything that might come between the commander and his ability to command his troops. The generals did not want to face the many problems entailed by creating a political commissar. The second thing that the Wehrmacht's experience showed was how important Hitler and his colleagues believed such a institution could be in maintaining control over the military at a time of mounting chaos.

German officers were only too aware of the Russian experience. They knew perfectly well that if most Nazis had their way, an institution similar to that utilized by the Russians would become a reality in the Wehrmacht. Many, if not most, Nazis were always suspicious of the army, and Hitler himself believed that his generals were always plotting against him. What better way to ensure that the generals behaved themselves than by introducing a "watchdog" to ensure that Germany's senior officers behaved? The problem, however, was that Hitler was dependent upon the army to achieve his long-term foreign policy goals. How could he hope to win a major war if military efficiency was undermined by these political commissars? Maybe because Hitler had experienced war firsthand, he understood that the generals had a point—too many cooks could spoil the broth.

Had the German Army been triumphant and had Germany won the war in 1943 or 1944, a good argument could be made that these leadership officers would have never been introduced in the Wehrmacht. Instead, it is possible that the Waffen SS would have gradually replaced the Wehrmacht. The former did not have NSFOs because every SS officer was expected to be a convinced Nazi. A special officer was not necessary to carry out these functions. In any case, the introduction of NSFO officers in the Wehrmacht was similar to the actions taken by the Soviets in the late 1930s and early 1940s. It was a temporary step aimed at stabilizing the military at a critical point in time.

Some General Observations

It is now time to ask the question "Is there anything we have learned from these seven cases that could be seen as general statements about the nature of civil-military relations in totalitarian states?"

The first observation that follows from this study is that value variance is a key variable affecting the nature of civil- military relations in a party-state. If a serious difference in values exists between the military and the political leadership and the latter has the option, it will take steps to bring these values into agreement by purging officers from the old regime and replacing them with officers who share its ideological or religious beliefs.

In many cases, however, it will not be possible to get rid of all potentially hostile officers. This is especially true if the regime faces a major threat or needs these officers to carry out its domestic or foreign policy goals. In such cases, the political leadership will have no alternative but to develop a series of political control devices to ensure that the military leadership will not be able to act against the interests of the party. In the end, regime survival is uppermost in the minds of the party leadership. This means compromising ideological purity, but the vast majority of such regimes will be prepared to pay this price.

At the same time, the political elite will begin a massive political socialization process to convince members of the military of the superiority of the new system when compared with the old one. The propaganda may be somewhat crude at times, but the regime will keep hammering home a message aimed at gaining acceptance of the new value structure.

Once these control devices—what Samuel Huntington called subjective control measures—are no longer necessary to ensure the loyalty and reliability of the army, they will not be eliminated. Rather, as is the case in many bureaucratic organizations, they will be transformed into what Huntington called objective control measures—institutions that serve to further the military's professionalism and autonomy by enabling the military to focus more closely on issues such as motivation, counseling, and changing the soldier's basic value system.[1]

As the transformation from subjective to objective control measures takes place, these former "control" officers will be coopted by the military. In this sense, the political officer will become similar to (but not the same as) the chaplain. Both serve two masters (the church or the party, as well as the military), but, as this study demonstrates, most of them looked upon themselves as military officers first and representatives of the party or church second. A chaplain's break with his sponsoring church could get him fired. Similarly, open defiance of the party, as was the case with several East German political officers who opposed the 1968 invasion of Czechoslovakia, could and did get one thrown out of the armed forces. However, it was the military and its strict hierarchical structure that determined the officer's future, something the young officer recognized very quickly.

Similarly, the nature of the political socialization process changes as the move from subjective to objective control measures takes place. Instead of the effort to bring about a massive "value change," the approach will more closely resemble what might be called a system maintenance program. The purpose will be to build on the ideas inculcated at an earlier period in order to ensure that the soldiers and sailors remained committed to the existing regime and willing to do its bidding.

It is also clear that in a party-state, the military operates best when it is isolated from society during periods of crisis and instability. The more contact soldiers have with civilians, the greater the likelihood that soldiers will pick up many of the civilians' attitudes—some of which may not support the regime. This was evident in the French Revolution, with the Russian (Imperial) Army, as well as with the Soviet military at the end of the USSR and the East German military at the end of the GDR.[2]

It is also worth noting that in contrast to the views of many Westerners, the party organization, including political officers, played a very important role in assuring the military's cohesion. This became especially clear in the closing days of a multi-ethnic state like the Soviet Union, when the country's military leaders resisted efforts to eliminate the special position the Communist Party enjoyed in that country. The same could be said for the East German party apparatus. The problem in that instance was that East Berlin was too confused and too chaotic to provide the political organs with the kind of guidance they needed to help maintain cohesion in the NVA. Left without centralized guidance, the political organs stood by helplessly, unable to perform the function they had carried out for many years.

The intensity of politicization in a party-state depends on a number of factors. First, as noted earlier, the degree of value divergence between the regime and the military is a key factor. The greater the difference in values, the more intense the political socialization and value modification will be. Second, the greater the degree of systemic instability, the harder the regime will push political socialization. This is what happened in the French and Soviet/Russian cases, as well as in the German Wehrmacht during the last two years of the war. This strategy was also attempted in the waning years of the GDR, although the lack of guidance from Berlin made such efforts ineffective.

Most important, for the purposes of this study, the decision to change from commissars to political officers indicates that the regime believes that it has gained the loyalty of the armed forces—at least in the realm of values. Political authorities may decide to declare war on the military—as happened under Stalin in Russia—but that was part of a larger political battle and led to the reintroduction of the political commissar, even if in a modified form. Once the purge was over and Stalin was sure of the army's loyalty, he went back to the more efficient political officer system. In this sense, close attention to these types of political control mechanisms can be important indicators of civil-military relations. Furthermore, given the more efficient role the political officer plays within the military, it should also come as no surprise that a military that is staffed by political officers will be more effective in carrying out its missions since it will not be afflicted by the inefficiencies of the commissar system.

This study also suggests that despite their strong commitment to their ideological beliefs, successful revolutionary leaders have been prepared to compromise when dealing with their military forces. This was especially evident in the Soviet case, where Lenin accepted the presence of military specialists despite open opposition on the part of many of his party colleagues. He made it clear on several occasions that the Bolsheviks could not have won the civil war without the assistance of the military specialists. However, he also was a strong supporter of subjective control measures to ensure their loyalty. The same could be said of the early days of the East German regime. As far as Hitler was concerned, he was realistic enough to know that a frontal attack on the military would work at cross purposes with his goal of restoring German power in Europe.

Concluding Thoughts

Despite the attention that political officers received during the Cold War, when they were depicted as an institution that was unique to a totalitarian state such as the former Soviet Union, an analysis of the roles played by chaplains in Cromwell's army and in the American military suggests that when it came to motivation and counseling, political officers carried out jobs that were part and parcel of any army.

Many years ago the Russian revolutionary and philosopher Michael Bakunin is reported to have said that "nothing unites the Russians like the Germans. If they did not exist, we would have to invent them." A somewhat similar comment could be made about political officers and chaplains. Despite their obvious differences, they both fulfilled very important functions as motivators, morale officers, and political socializers. We might decide to call either of these officers something else, but the fact is that every army—whether democratic or totalitarian—needs specially trained officers to perform these crucial tasks. Indeed, I cannot think of anything that more accurately demonstrates just how related the functions of these two institutions are than what has happened in the postcommunist militaries.

In every case I am aware of, when political officers were abolished, chaplains were introduced. This has occurred in the Czech Republic, the Ukraine, Hungary, Georgia, and, most important, Russia.[3] Insofar as the Russian military is concerned, Moscow reached an exclusive agreement with the Russian Orthodox Church in 1999 to provide chaplains for the armed forces. One source noted in this regard that more than "100 Russian Orthodox churches and chapels are operating on military bases throughout the country."[4] It is not clear to what degree these priests operate as counselors or deal with morale problems. While they are probably not in the chain of command the way a political officer was, it is reasonable to suspect that they are carrying out many of the same duties that chaplains did in the American armed forces. The temptation for a line officer to send a soldier with a personal problem to the chaplain can be overwhelming. If the chaplain is there, the military bureaucracy will find a way to make use of him. In addition, the Kremlin has introduced education (Vospitalne) officers into the postcommunist Russian Army—and these officers deal with things directly related to political socialization, motivation, and morale.

Even more interesting has been the movement to get chaplains from different militaries to work together in dealing with issues such as Bosnia. In March 1999, for example, chaplains from thirty-three countries met outside Vienna, Austria, to consider what they could do to increase cooperation. In the past, they had no role in NATO deliberations "but simply served with their nation's troops during exercises." The idea was to set up a NATO council dealing with cooperation between the chaplains of all sixteen NATO countries. Its goal would be to advise "commanders on religious sensitivity issues, working with Partnership for Peace nations struggling with the idea of democracy and human rights, helping each other understand how we can work together in our multinational operations."[5] This is something new, a form of military cooperation that was unknown in the past.

During the Cold War, many in Washington assumed that the political officer in the Soviet military was really a "commissar," which by definition suggested that more conflict existed within the Soviet armed forces than was the case. The assumption was that the Soviet military resented the party and was locked in a never-ending struggle, as the military fought for increased autonomy from a party that was attempting to penetrate it at every opportunity. Needless to say, this had important implications for our intelligence estimates. We saw these individuals as a negative force, when they were really playing a positive role. The result was a skewed analysis—if the Soviet military had been forced to fight, it might have been more potent than many Westerners projected. The same would have been true, of course, when some people occasionally assumed that no differences existed between the generals and the party leadership. There were important differences, but too often we pretended that they did not exist. Needless to say, such misperceptions had important policy implications.

Looking at the role a morale or motivational officer plays in any armed forces is not a panacea to understanding civil-military relations in a single party-state. It does, however, help us understand one crucial aspect of the overall relationship. Assuming that we are not dealing with a state run by the military, if control devices are lacking, it probably means that there is agreement on values, an important starting point for our understanding civil-military relations in such systems.

The same could be said of the importance of looking at the actions of political officers—or, for that matter, of chaplains. When these "social" officers get involved in a new and different area, it generally signifies that the armed forces are experiencing new and different problems, regardless of whether these are racism, alcoholism, drugs, AIDs, or so on. Understanding this aspect of "social" officers' jobs is key in making an overall evaluation of a military, of its stature and place in its society.

Heraclitus was right: nothing endures but change. If a model cannot account for change, it has very limited value. Such paradigms become sterile, stagnant, one dimensional, and ultimately useless. In short, a model must be able to deal with the complex and often frustrating world of change if it is to be useful as an analytical device.

What I have attempted in this book is to suggest that the kinds of control devices adopted by party-states or totalitarian regimes will change over time. This hardly qualifies as a grand theory, but it does point out the pitfalls of assuming that conflict is ubiquitous in civil-military relations or that the interests of the party leadership and those of the military are always identical. At one point, relations between the two may conflict, while at another time they could well be in strong agreement. If we are going to deal successfully with such regimes in the future, it is crucial that we understand the nature of the relationship between the armed forces and the party.

This purpose of this study was to construct some generalized statements relative to civil-military relations in party-states. Even in this case, however, my goal was rather modest since it focused on only one aspect of civil-military relations in such polities by looking at Huntington's subjective control devices and asking the question, Could they be transformed into objective control devices over time, and if so, what are the implications for our understanding of politics in such systems?

In making such comments, I am clearly on the opposite side of those who believe that we are in a position now to conceptualize civil-military relations in a political system such as Russia. I would argue that we are far from understanding the situation in one country, let alone across state boundaries. I do not mean to suggest that such analyses are not useful. Indeed, this study attempts to be comparative. I do argue, however, that we do not yet have a conceptual framework that explains change, as well as the many variables that influence civil-military relations in any polity. We only delude ourselves if we believe that we have found a shortcut, a model or framework that does not require knowledge of culture, language, and history.[6]

It is my view that political scientists must engage in theory-building by working from the bottom up. By this, I mean that we are faced with the prospect of systematically examining a number of case studies in the hope that we will be able to extract some general principles that over time will enable us to build and test more linked relationships. While I did not intend to explain civil-military relations on a macro level in this book, I did try to make some universal statements about the nature of one aspect of civil-military relations in party-states and how these relationships have led to changes in some of the structures associated with civil-military relations in these types of systems.

We must learn to crawl before we can hope to run. Unfortunately, there are no shortcuts in this process. Knowledge of other languages, cultures, history, and political systems is crucial. Only by taking a careful, in-depth look at a number of different political systems will we begin to understand how civil-military relations—and politics in general—work in a broader context.

Notes

1. Samuel Huntington's comments on the relationship of objective and subjective control measures are contained in his now classic work *The Soldier and the State* (New York: Vintage, 1957), 83.

2. This was especially true in the former GDR. See my *Requiem for an Army: The Demise of the East German Army* (Boulder, Colo.: Rowman & Littlefield, 1998).

3. See, for example, "Chaplains Reinstated," *Times Picayune,* 14 December 1993; "Ukraine to Accept Rabbis as Military Chaplains," *The Jerusalem Post,* 23 December 1992; "Chaplains Will Replace Political Officers in the Georgian Army," *Defense and Security,* 31 March 2000; "Taking Faith to the 'New' Front Lines," *Christian Science Monitor,* 4 March 1999; "New on the Front Lines: Army Chaplains," *New York Times International,* 15 January 2000; and Andrei Zolotov, Jr., "Army Cautiously Welcomes Back the Church in 'Near Abroad,'" *Moscow Times,* 5 May 1999.

4. RFE/RL Daily Report, 5 May 1999.

5. "Taking the Faith to the 'New' Front Lines."

6. Attempts to apply a general theory, in the form of the "rational actor model" by John Williams and Mark Webber, contribute little to our understanding of Russian civil-military relations. The authors come up with little that is substantive or new to a seasoned analyst of Russian military affairs. See John H. P. Williams and Mark J. Webber, "Evolving Russian Civil-Military Relations: A Rational Actor Analysis," *International Transactions* 24, no. 2, 115–130.

Index

About the Author

Dale R. Herspring, a senior professor in the department of political science at Kansas State University, is a retired U.S. diplomat and naval officer. He is the author of eight books and more than sixty articles dealing with German, East European, and Russian/Soviet civil-military relations. His most recent book, *Requiem for an Army: the Demise of the East German Military,* was translated into German. He is now working on a book with Hans-Werner Weber to be entitled, *Rainer Eppelmann and the Politics of Human Rights in Germany.*